Mass Transfer Driven Evaporation from Capillary Porous Media

Mass Transfer Driven Evaporation from Capillary Porous Media offers a comprehensive review of mass transfer driven drying processes in capillary porous media, including pore-scale and macro-scale experiments and models. It covers kinetics of drying of a single pore, pore-scale experiments and models, macro-scale experiments and models, and understanding of the continuum model from pore-scale studies.

The book:

- Explains the detailed transport processes in porous media during drying.
- Introduces cutting-edge visualization experiments of drying in porous media.
- Describes the pore network models of drying in porous media.
- Discusses the continuum models of drying in porous media based on pore-scale studies.
- Points out future research opportunities.

Aimed at researchers, students, and practicing engineers, this work provides vital fundamental and applied information to those working in drying technology, food processes, applied energy, and mechanical and chemical engineering.

Advances in Drying Science and Technology

Series Editor:
Arun S. Mujumdar
McGill University
Quebec, Canada

Industrial Heat Pump-Assisted Wood Drying
Vasile Minea

Drying of Biomass, Biosolids, and Coal: For Efficient Energy Supply and Environmental Benefits
Shusheng Pang, Sankar Bhattacharya, Junjie Yan

Drying and Roasting of Cocoa and Coffee
Ching Lik Hii and Flavio Meira Borem

Heat and Mass Transfer in Drying of Porous Media
Peng Xu, Agus P. Sasmito, and Arun S. Mujumdar

Freeze Drying of Pharmaceutical Products
Davide Fissore, Roberto Pisano, and Antonello Barresi

Frontiers in Spray Drying
Nan Fu, Jie Xiao, Meng Wai Woo, Xiao Dong Chen

Drying in the Dairy Industry
Cécile Le Floch-Fouere, Pierre Schuck, Gaëlle Tanguy, Luca Lanotte, Romain Jeantet

Spray Drying Encapsulation of Bioactive Materials
Seid Mahdi Jafari and Ali Rashidinejad

Flame Spray Drying: Equipment, Mechanism, and Perspectives
Mariia Sobulska and Ireneusz Zbicinski

Advanced Micro-Level Experimental Techniques for Food Drying and Processing Applications
Azharul Karim, Sabrina Fawzia and Mohammad Mahbubur Rahman

Mass Transfer Driven Evaporation from Capillary Porous Media
Rui Wu, Marc Prat

For more information about this series, please visit: www.routledge.com/Advances-in-Drying-Science-and-Technology/book-series/CRCADVSCITEC

Advances in Drying Science and Technology

Series Editor:
Dr. Arun S. Mujumdar

It is well known that the unit operation of drying is a highly energy-intensive operation encountered in diverse industrial sectors ranging from agricultural processing, ceramics, chemicals, minerals processing, pulp and paper, pharmaceuticals, coal polymer, food, forest products industries as well as waste management. Drying also determines the quality of the final dried products. The need to make drying technologies sustainable and cost effective via application of modern scientific techniques is the goal of academic as well as industrial R&D activities around the world.

Drying is a truly multi- and interdisciplinary area. Over the last four decades, the scientific and technical literature on drying has seen exponential growth. The continuously rising interest in this field is also evident from the success of numerous international conferences devoted to drying science and technology.

The establishment of this new series of books entitled *Advances in Drying Science and Technology* is designed to provide authoritative and critical reviews and monographs focusing on current developments as well as future needs. It is expected that books in this series will be valuable to academic researchers as well as industry personnel involved in any aspect of drying and dewatering.

The series will also encompass themes and topics closely associated with drying operations, e.g., mechanical dewatering, energy savings in drying, environmental aspects, life cycle analysis, technoeconomics of drying, electrotechnologies, control and safety aspects, and so on.

About the Series Editor

Dr. Arun S. Mujumdar is an internationally acclaimed expert in drying science and technologies. He is the Founding Chair in 1978 of the International Drying Symposium (IDS) series and Editor-in-Chief of *Drying Technology: An International Journal* since 1988. The fourth enhanced edition of his *Handbook of Industrial Drying* published by CRC Press has just appeared. He is recipient of numerous international awards including honorary doctorates from Lodz Technical University, Poland and University of Lyon, France.

Please visit www.arunmujumdar.com for further details.

Mass Transfer Driven Evaporation from Capillary Porous Media

Edited by
Rui Wu and Marc Prat

CRC Press
Taylor & Francis Group
Boca Raton London New York

CRC Press is an imprint of the
Taylor & Francis Group, an **Informa** business

First edition published 2023
by CRC Press
6000 Broken Sound Parkway NW, Suite 300, Boca Raton, FL 33487-2742

and by CRC Press
4 Park Square, Milton Park, Abingdon, Oxon, OX14 4RN

CRC Press is an imprint of Taylor & Francis Group, LLC

ISBN: 9780367416850 (hbk)
ISBN: 9781032362311 (pbk)
ISBN: 9781003011811 (ebk)

DOI: 10.1201/9781003011811

Typeset in Times
by codeMantra

Contents

Preface

The present book is dedicated to evaporation from a porous medium. For geoscientists, this is a fundamental process occurring in the vadose zone. In engineering, evaporation is at the core of the drying process, i.e. the process of removing moisture from solids via a liquid–vapor phase change. A drying step is encountered in the manufacturing or packaging process of many products. This concerns almost every food or non-food product used in solid form. Evaporation in a porous medium is also a key aspect in many other applications such as the recovery of light oils; the decontamination of soils contaminated by light hydrocarbons; the alteration of rocks and building materials due to salt crystallization; the injection of CO_2 in saline aquifers; and the operation of many systems, such as PEM fuel cells, air coolers, porous evaporators, and distillation units.

The significance of the numerous applications where evaporation in a porous medium is a central process has motivated many studies and modeling efforts. Over a period spanning over about 60 years, from the thirties to the late eighties, this has led to the elaboration of models within the framework of the continuum approach to porous media. These models are still widely used in geosciences and engineering because they are relatively easy to solve via standard numerical discretization techniques and well adapted to the simulation of evaporation in porous systems containing many pores, which is the most common situation in the applications. They are formulated in terms of standard partial differential conservation equations and therefore can also be relatively easily implemented in commercial software. Although these models are quite useful, they suffer from important shortcomings. More specifically, the prediction of the evaporation rate variation during drying, and thus of the drying time, is still difficult because the drying or evaporation theory lacks a solid formulation of the coupling conditions between the transfers in the porous medium and in the external gas, a problem more generally referred to as a porous medium/ free flow problem. Also, the effect of capillary liquid films and other secondary capillary structures, such as liquid bridges and capillary rings, is somewhat hidden in the macroscopic parameters, whereas it becomes more and more clear that it would be more interesting to consider them explicitly in the continuum model formulation.

These shortcomings are one motivation for developing alternative approaches so as to be in a position to possibly improve the prediction capabilities of continuum models. In fact, the continuum models provide little information on the physics at pore scale, whereas it seems obvious that a thorough understanding and description of the phenomena occurring at pore scale and at the scale of the pore network should be the fundamental basis for developing improved continuum models. Although the possible improvement of continuum models is a strong motivation, the pore scale phenomena are also worth studying independently of continuum models. For example, a detailed understanding of the pore scale mechanisms can help interpreting and analyzing experimental observations. Also, thin porous systems, such as papers, textiles, and the ones encountered in PEM fuel cells, for instance, do not meet the usual length scale separation criterion at the core of the classical continuum models and require therefore alternative approaches. Also, it is not easy to understand the impact of a porous microstructure on the drying kinetics from continuum models. To this end, again, alternative approaches are desirable.

In fact, in terms of modeling, two main alternative approaches have been developed since the early nineties: pore network models (PNMs) and direct simulations. In the case of drying, the latter mainly refer to simulations based on Lattice Boltzmann models (LBMs). In this type of simulations, evaporation and the resulting evolution of the liquid–gas distribution in the porous medium are simulated directly at the pore scale and can be performed in principle over the complex three-dimensional pore space of real porous microstructures that can be obtained by modern imagery techniques. However, the use of LBM simulations for the study of evaporation in porous media has been less developed so far than pore network simulations. It can be surmised that this situation will change rapidly in the future since LBM codes become increasingly available and accessible. The more intensive use of PNM can be explained by the fact that the computational costs of PNM simulations are much less. Also, simpler drying PNMs are based, in part, on standard finite volume or finite difference methods and relatively simple algorithms, which make the development of a PNM code on simple lattices reasonably straightforward. Drying PNMs have been developed and used over the last 30 years by several groups over the world. They have led to numerous results and insights, offering a rather detailed understanding and description of the physics of drying at the pore network scale. However, contrary to LBM simulations, PNMs are based on a geometrical simplification of the pore space, which sometimes can be a shortcoming. Also, it can be noted that drying PNM simulations have been performed so far mostly on model lattices, such as square or cubic lattices. The important step of developing drying PNM simulations on pore networks extracted from microstructure 3D digital images is still largely a prospect.

On the experimental side, a major evolution since the end of eighties is the use of microfluidic devices, also referred to as micromodels. Experiments in these systems have greatly helped to improve our understanding of the evaporation process, often in relation to PNM simulations. The use of NMR for determining the saturation profiles in standard column experiments is also a major evolution.

In this context, the objective of the book is to illustrate the above-summarized evolution in the study of evaporation in porous media. As a matter of fact, covering all the aspects cannot be made in a single book. For this reason, the more specific objective of the book is to address the fundamental situation where drying is controlled by mass transfer, i.e. sufficiently slow for the temperature variations to be negligible. Also, the focus is on the capillary porous media, i.e. the category of porous media for which sorption phenomena and the Kelvin effect can be ignored. Furthermore, mechanical effects such as porous matrix deformation or fracturing are not considered. The emphasis is on the fundamental mechanisms controlling the evolution of the liquid-phase distribution and the evaporation rate in capillary porous media and the modeling tools enabling one to compute them.

The book contains seven chapters, namely:

Chapter 3: Pore Scale Experiments on Evaporation from Porous Media
Chapter 4: A Mesoscopic Approach for Evaporation in Capillary Porous
    Media: Shan Chen Lattice Boltzmann Method
Chapter 5: Pore Network Models for Evaporation in Porous Media
Chapter 6: Continuum Models
Chapter 7: A Continuum Approach to the Drying of Small Pore Networks

Chapter 1 proposes a state of the art. The three main modeling approaches mentioned above are discussed with an emphasis on the LBM and PNM approaches. The
review of the literature regarding the continuum models is presented in more detail
in Chapter 6. The significance of single tube and microfluidic device experiments is
illustrated with a presentation of key results. Research needs are discussed.

Chapter 2 discusses evaporation in the most elementary system: a single capillary tube. A capillary tube can be seen as an elementary component of pore network
models. The most striking result discussed in this chapter is the impact of the tube
cross-section geometry. When the cross section is polygonal, the evaporation kinetics
can be much faster than in a circular tube due to the impact of the flow in the corner
films trapped by capillarity within the tube wedges. The dynamics of the corner
films is controlled by the competition between viscous, gravity, and capillary forces,
a common feature with evaporation in a capillary porous medium.

Chapter 3 well illustrates the significance of evaporation experiments in a microfluidic device. It is shown that the gas invasion process in the pore network resulting
from evaporation can be subtler than that considered in the standard invasion percolation process, generally considered as being the relevant process for describing
the evolution of the liquid–gas distribution when capillary forces are dominant. It is
shown that the invasion is controlled not only by the constrictions of the pore space
but also by the invasion in the larger cavities of the pore space, referred to as pore
bodies, because of the sudden geometrical expansion at the junction between a constriction, also referred to as a pore throat, and a pore body. This leads to identify a
specific effect, referred to as the capillary valve effect, affecting the pore body invasion. This effect leads to an overlap between the distribution of the invasion capillary
pressure thresholds of the pore throats and the one for the pore bodies. The rupture of
corner films due to a geometrical effect is also analyzed as well as capillary instabilities leading to the possible refilling of pores by the liquid.

The potentialities of LBM simulations are illustrated in Chapter 4. The chapter
well illustrates several important features of evaporation at the pore scale such as
the capillary pumping effect, the preferential emptying of largest pores, the abrupt
changes in the liquid–gas interface position, and the Haines jumps or the capillary
instabilities leading to refilling of pores.

Chapter 5 presents some recent works based on pore network models for drying
considering the capillary valve effect, continuous and discontinuous corner liquid
films, and capillary instability effects. The results of the pore network models are
validated from comparisons with the experimental data presented in Chapter 3.

Continuum models are presented and discussed in Chapter 6. Two main classes
of continuum models are distinguished: the models based on the local equilibrium
(LE) assumption and the non-local equilibrium (NLE) models. In the context of

continuum models, the equilibrium in question must be understood at the scale of the representative elementary volume (REV). The most classical models are based on the LE assumption. This means that the vapor pressure within the REV in the presence of liquid within the REV is assumed to be equal to the one prevailing at the menisci within the REV, whereas the average vapor pressure over the REV is considered to be less in the non-equilibrium process corresponding to evaporation with the NLE models. The emphasis is also put on the interfacial coupling conditions with the mass transfer in the external gas with a review of the main formulations used in the literature. It is argued that the problem of the interfacial coupling condition cannot be simply considered as a standard conjugate problem.

Chapter 7 is an illustration of the use of PNM simulations in relation to continuum models. Because of the rather small networks considered, characterized by an obvious lack of length scale separation, the comparison leads to several original features. Contrary to the classical continuum models for which parameters only depend on the local saturation, macroscopic parameters depend here on both the local saturation and the overall saturation. This introduction of several sets of parameters can be interpreted as a memory effect due to the lack of length separation. Also, some internal parameters depend on the boundary condition, which can be seen again as a consequence of the lack of length scale separation. This chapter can also be seen as an attempt for modeling the drying process in a thin system using a continuum approach.

These chapters have been written by different contributors. As a result, one should accept some variety in the writing style and the notations. The priority was in the representativeness of the presented material compared to the state of the art. We are grateful to the contributors. We would also like to express our acknowledgments to the team of Taylor & Francis for its support and patience since it always takes time to gather contributions from various people.

Last but not least, we, of course, hope that the book will be of interest to many people, from newcomers in the field to experts, from academia to industry.

January 2022
Marc Prat
Rui Wu

Editors

Rui Wu is an Associate Professor at Shanghai Jiao Tong University. He serves as the deputy director of the Institute of Thermophysics Engineering. He earned his PhD degree from Chongqing University in 2012. In 2013, he became an Alexander von Humboldt Fellow. He worked in the Chair of Thermal Process at Otto-von-Guericke University from 2014 to 2015. His research interests include two-phase transport in porous media and interfacial phenomena.

Marc Prat is CNRS Senior scientist at the Fluid Mechanics Institute in Toulouse, France. He is Associate Editor of the *Journal of Porous Media* and the author or co-author of more than 150 articles in scientific archival journals. He has been the supervisor or co-supervisor of more than 50 PhD students and P.I. in many projects funded by various industrial partners, national or international agencies. His main research interest is the study of transport phenomena in porous media in relation to engineering applications, with a special interest in the situations involving phase change phenomena.

Contributors

Supriya Bhaskaran
Birla Institute of Technology and
  Science
Pilani–Hyderabad Campus, India
and
Otto-von-Guericke-Universitat
  Magdeburg
Magdeburg, Germany

Abdolreza Kharaghani
Otto-von-Guericke-Universität
  Magdeburg
Magdeburg, Germany

Xiang Lu
Otto-von-Guericke-Universität
  Magdeburg
Magdeburg, Germany

Shubhani Paliwal
Birla Institute of Technology and
  Science
Pilani–Hyderabad Campus, India
and
Imperial College London
London, United Kingdom

Debashis Panda
Birla Institute of Technology and
  Science
Pilani–Hyderabad Campus, India
and
Imperial College London
London, United Kingdom

Marc Prat
Institut de Mécanique des Fluides de
  Toulouse
Toulouse, France

Vikranth Kumar Surasani
Birla Institute of Technology and
  Science
Pilani–Hyderabad Campus, India

Evangelos Tsotsas
Otto-von-Guericke-Universität
  Magdeburg
Magdeburg, Germany

Rui Wu
Shanghai Jiao Tong University
Shanghai, China

Githin Tom Zachariah
Bernal Institute
University of Limerick
Limerick, Ireland

1 Slow Evaporation in a Capillary Porous Medium
A State of the Art

Marc Prat
Institut de Mécanique des Fluides de Toulouse

CONTENTS

1.1 INTRODUCTION

Evaporation from porous media is a key process in numerous important applications. This is a very important process in geosciences in relation to the phenomena occurring in the critical zone (Or et al., 2013) and the meteorology. Related issues are the soil salinization processes (Hassani et al., 2021) or the alteration of rocks via evaporation cycles (Benavente et al., 2011). A drying step is commonly encountered in many engineering processes (Mujumdar, 2006) involving food products or non-food products such as wood, papers, washing powders, etc. This is also an important process in the decontamination of soils contaminated by light hydrocarbons (Shah et al., 1995) or the underground storage of CO_2 in aquifer (Muller et al., 2009). Air fresheners, evaporative coolers (Wang et al., 2022), water transfer in fuel cells (Zenyuk et al., 2016) and water transfer in building walls (Fang et al., 2021) are other examples, to name only a few, where evaporation in a porous medium is a key aspect. As a result, it is not surprising that the topic has been the subject of many studies over the years and is still a very active research topic. In most situations, the evaporation process is limited by transport phenomena; it is not controlled by the evaporation process itself, i.e. the liquid–vapor phase change occurring at the interface between the liquid and the gas (Schrage, 1953). The transport phenomena in question are the mass transfer (the vapor forming at the liquid–gas interface must be transported away from the interface) and the heat transfer (evaporation is an endothermic process, and therefore,

DOI: 10.1201/9781003011811-1

heat should be supplied to the liquid–gas interface for obtaining high evaporation rates). As a result, in many situations, evaporation depends on both the mass and heat transfers occurring in the liquid and gas phases (and the solid phase as regards the heat transfer). However, when the evaporation rate is sufficiently small, the energetic aspect can be neglected because the associated temperature variations are small, and evaporation is then driven essentially by the mass transfer. A typical example is the convective or diffusive evaporation of water at room temperature. This allows focusing on the evaporation process and the evolution of the phase distributions in the porous medium without the additional complexity linked to the coupling with the heat transfer. In other words, the temperature can be considered as spatially uniform in this case and constant. This situation is also referred to as quasi-isothermal drying or evaporation, and also as slow evaporation (or slow drying) since the evaporation rate must be sufficiently low for the temperature variations to be negligible. In this book, the focus is on quasi-isothermal evaporation unless otherwise mentioned. A strong motivation for this simplification lies in the fact that the evolution of the phase distribution during drying or evaporation is a key aspect. In an approach of ascending order of complexity, this evolution should be well understood first in the case where temperature variations can be ignored.

Another important distinction concerns the type of porous media. It is convenient to distinguish the capillary porous media from the hygroscopic porous media. The former are characterized by the fact that the volume of liquid that can be fixed within the pore space by adsorption phenomena (Adamson, 1990) represents a negligible fraction of the pore space. Hence, the liquid is essentially held by capillarity. This typically corresponds to porous media with pore size greater than about 1 μm. The random packings of sub-millimetric particles used in many experiments are archetypical examples of capillary porous media. By contrast, the hygroscopic porous media, such as wood or concrete, are characterized by much smaller pores in the sub-micronic range, and a significant fraction of liquid can be fixed by adsorption in the pore space. Also, the impact of the meniscus curvature on the thermodynamic equilibrium, i.e. the Kelvin effect (Adamson, 1990), is negligible in capillary porous media but an important effect in hygroscopic porous media. Moreover, the liquid transport in the bound liquid is considered as an important phenomenon in hygroscopic porous media, whereas it is considered as negligible in capillary porous media. Then again according to an approach of ascending order of complexity, it makes sense to try first to understand, describe and model the presumably simpler case of capillary porous media. For this reason, capillary porous media are the main target in this book unless otherwise mentioned.

As for the study of other phenomena in porous media, various experimental, numerical and theoretical approaches have been used or developed for studying evaporation in capillary porous media. The remaining of the present chapter proposes an overview of the corresponding works in relation to the material in the other chapters of the book.

Basic results on drying of capillary porous media are recalled in Section 1.2. Experiments with microfluidic devices (micromodels) are discussed in Section 1.3. Modeling approaches are presented in Section 1.4. An overview of research needs is presented in Section 1.5. A summary is given in Section 1.6.

1.2   SLOW DRYING OF CAPILLARY POROUS MEDIA

In basic research on drying, the archetypical situation exemplified in Figure 1.1 is often considered. The capillary porous medium is sealed on each side except on the top surface which is in contact with the external gas (humid air considered as a binary mixture of air and water vapor). The porous medium is considered as non-deformable and saturated by liquid, water in most cases, when the experiment starts.

An air flow can be imposed over the top surface of the porous medium (convective drying) but experiments can also be performed without a forced convection flow. In both cases, i.e. with or without forced air flow, the vapor partial pressure at the porous medium surface (equal to the saturated vapor pressure at the very beginning of the experiment) is greater than the vapor pressure in the surrounding air. This vapor pressure difference induces a mass transfer at the top surface of the porous medium. Hence, $j \propto p_{vi} - p_{v\infty}$, where j is the evaporation flux, and p_{vi} and $p_{v\infty}$ are the vapor pressure at the top surface of the porous medium and in the surrounding air, respectively. As a result of this mass transfer, evaporation occurs and the gas phase gradually replaces the liquid in the pore space. This evolution is typically characterized by the measurement of the evolution of the evaporated mass m_e as a function of time (t) from which the drying kinetics, i.e. the variation of the evaporation rate $J = \dfrac{dm_e}{dt}$ as a function of the overall liquid saturation $S = \dfrac{m_T - m_e}{m_T}$, where m_T is the total mass of liquid saturating the pore space at the beginning of the experiment, can be obtained. Saturation S represents the fraction of the pore space volume occupied by the liquid phase. Experiments with capillary porous media typically lead to the result exemplified in Figure 1.2 leading to distinguish three main drying periods (Van Brakel, 1980): the constant drying period (CRP), the falling drying period (FRP) and the receding front period (RFP). The existence of the CRP is perhaps the

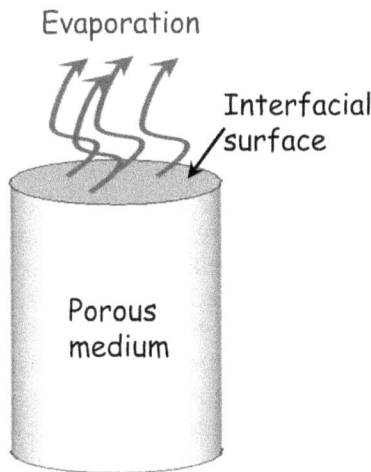

FIGURE 1.1   Archetypical drying situation with interfacial surface with the external transfers at the sample top.

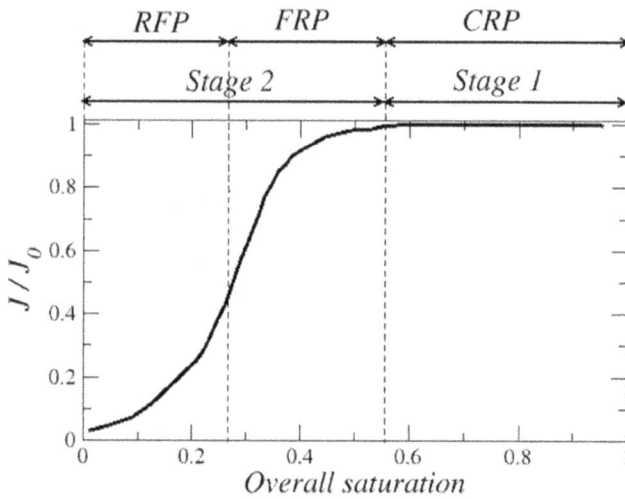

FIGURE 1.2 Typical slow drying kinetics of a capillary porous medium. J is the evaporation rate and J_0 the evaporation rate at the beginning of the experiment. (Adapted from Gupta et al. (2014).)

most striking feature making the drying of capillary porous media quite different from evaporation in a single circular capillary tube (Whitaker, 1991). The view is that liquid is sufficiently present at the porous medium surface for the evaporation rate not to be significantly affected over the CRP. Capillary effects are responsible for the extended liquid presence, pulling the liquid from the interior of the porous medium via the menisci in the smaller surface pores which are not yet invaded by the gas. As drying continues, the fraction of surface pores invaded by the gas increases, the liquid phase becomes more ramified (its effective permeability decreases) and the evaporation rate eventually starts decreasing. This corresponds to the transition between the CRP and the FRP in Figure 1.2. In the FRP, a dry zone eventually forms at the porous medium top. The gradual expansion of this dry zone corresponds to the RFP in Figure 1.2 (Van Brakel, 1980). As shown in Pel et al. (1996), the internal front position varies linearly with time in the RFP and not as the square root of time as in circular capillary tube (Whitaker, 1991). This is the consequence of two factors: the liquid phase is percolating (in the region below the front) and a viscous flow occurs in the liquid phase also during the RFP. This description of the drying kinetics in three main periods is sometimes simplified with the distinction of only two stages (Brutsaert and Chen, 1995): Stage 1 – evaporation corresponds to the CRP, whereas in Stage 2 –evaporation corresponds to the FRP and the RFP (Figure 1.2). It can also be noted that the evaporation rate is often not truly constant in the CRP. It simply varies much less than in Stage 2 – evaporation.

Explaining the CRP is one of the fundamental goals of the capillary porous media drying theory. Several factors have been discussed in this regard. On the side of the external mass transfer (the transfer in the surrounding gas from the porous medium surface), it has been shown that a partially wet surface could lead to about the same evaporation rate as if the surface was fully liquid, provided that the wet surface fraction was sufficiently high and the characteristic size of the external mass transfer layer (thickness of the mass boundary layer in convective drying, for instance) was sufficiently large compared to the pore size (Suzuki and Maeda, 1968; Schlünder, 1988; Lehmann and Or, 2013; Attari-Moghaddam et al., 2018; Talbi and Prat, 2019). Regarding the liquid transport in the porous medium, it has been shown (Coussot, 2000) that the internal transport was not limiting, i.e. the liquid could be supplied to the surface by the capillary effects, over a significant range of saturation provided that the initial evaporation rate was not too high. Also, liquid films in the region invaded by the gas phase can contribute to maintain a high vapor pressure at the surface via the liquid transport in the films (Chauvet et al., 2009; Yiotis et al., 2012). However, the detailed analysis of the transfer in the interfacial region (region where the top region of the porous medium is in contact with the surrounding air) is still a research topic (Talbi and Prat, 2021). In particular, the correct prediction of the FRP/CRP transition (or Stage 1–Stage 2 transition, Shokri and Or, 2011) remains a challenge.

In addition to the drying kinetics, which is, of course, of major interest since characterizing the drying time is of primary interest, the evolution of the liquid distribution in the porous medium is also of great interest. Obtaining this information is more challenging. Values of the saturation in horizontal slices can be obtained using scanning neutron radiography (Pel et al., 1993) or NMR (Pel et al., 1996; Thiery et al., 2017). When the drying is sufficiently slow, saturation profiles such as the ones depicted in Figure 1.3 are typically obtained. As can be seen, the saturation profiles are essentially flat (neglecting the fluctuation around the mean value that is probably due to the fact that the porous medium is not perfectly homogeneous in this example) over the CRP (Stage 1).

The drying regime illustrated in Figure 1.3 with flat saturation profiles during most of the CRP corresponds to the capillary regime (Prat, 2002). Under the same drying conditions, i.e. initial evaporation rate and sample dimensions, non-flat profiles with the slice saturation increasing with the depth in the porous medium can be also observed when the viscous forces (samples with sufficiently small pores) or the gravity forces (samples with sufficiently coarse grains, for instance) are not negligible compared to the capillary forces. The corresponding regimes are referred to

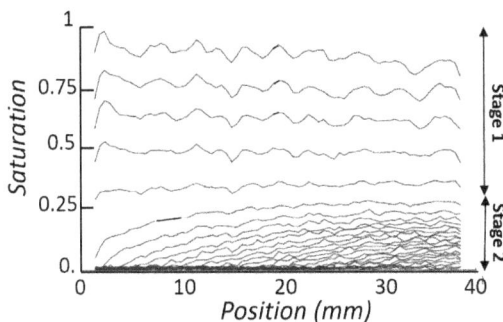

FIGURE 1.3   Typical saturation profiles during drying in a capillary porous medium in the capillary regime. (Adapted from Gupta et al. (2014).)

as the capillary–viscous and capillary–gravity regimes, respectively. In addition to usual dimensionless numbers such as the capillary number or the Bond number for characterizing the competition between capillary forces and viscous or gravity forces (Prat, 2002), characteristic length scales can be associated with the capillary–viscous and capillary–gravity regimes (Lehmann et al., 2008). These length scales actually characterize the size of the two-phase zone which can develop in the sample (Shaw, 1987; Prat and Bouleux, 1999). When these lengths are significantly greater than the sample height, the capillary regime is expected.

The saturation profiles, such as the ones depicted in Figure 1.3, and the drying kinetics (Figure 1.2) are macroscopic data and main objectives for the drying models. However, this type of data provides little information on what happens at the pore scale and within the pore space. In this regard, experiments with quasi-2D model porous media have been instructive as discussed in the next section.

1.3   MICROFLUIDIC EXPERIMENTS

The first experiment of this type is perhaps the one presented in Shaw (1987) in which a packing of small spheres was confined between two glass slides. This quasi-2D experiment corresponds to the capillary–viscous regime. The size of the two-phase zone was shown to vary with the capillary number according to a power law involving a non-trivial exponent. In fact, it is now well established that this regime can be analyzed in the framework of the percolation theory (Stauffer and Aharony, 1992), more precisely the theory of percolation in a gradient (e.g. Tsimpanogiannis et al., 1999; Prat and Bouleux, 1999 and references therein). This theory enables one to express the power law exponent as a function of critical exponents of the percolation theory (Prat and Bouleux, 1999). In other words, the study by Shaw can be seen as the first indication that the study of drying could be addressed within the framework of percolation theory and its variants. These studies made also clear that drying in 2D systems is different from drying in 3D pore space because of the fundamentally different percolation properties between a 2D and a 3D system. In particular, the extent of the two-phase zone is controlled by a non-trivial exponent in 2D as mentioned above, whereas the extent of the two phase zone trivially varies as the inverse of the capillary number (capillary–viscous regime) or the Bond number (capillary–gravity regime) in 3D (exponent 1) except for a tiny zone of negligible extension, in general. Sometimes, this had led to confusion and even to erroneous papers!

Almost 10 years later, experiments in a 2D network of etched channels (Laurindo and Prat, 1996) made clear that slow drying in this system led to fluid patterns very similar to the drainage capillary fingering pattern (Lenormand et al., 1988). The latter corresponds to a variant of percolation theory referred to as invasion percolation (Wilkinson and Willemsen, 1983). As illustrated in Figure 1.4, the invasion percolation process is characterized by the formation of numerous liquid clusters, i.e. the fragmentation of the liquid phase.

Another experiment on a model system performed in the nineties (Engøy et al., 1991) well illustrates a key concept in the analysis of drying: the so-called capillary pumping effect. Consider the system illustrated in Figure 1.5 formed by two interconnected channels of different widths.

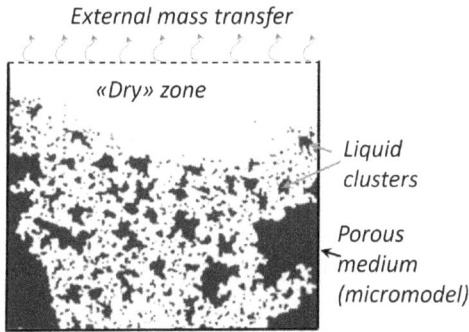

FIGURE 1.4 Liquid (in black) and gas (in white) distribution in a micromodel, a square network of interconnect channels of rectangular cross-section. Liquid films are present in the gas pores, including in a fraction of the "dry" zone. (Adapted from Laurindo and Prat (1996).)

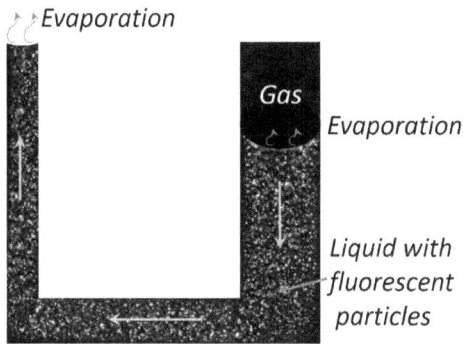

FIGURE 1.5 Evaporation in a system of two interconnected channels of different sizes. The meniscus recedes in the largest channel due to the capillary pumping effect by the meniscus at the entrance of the smaller channel. (Adapted from E. Ghiringhelli PhD thesis.)

As illustrated in Figure 1.5, the key observation is that the meniscus recedes in the largest channel, whereas it stays stuck at the channel entrance in the smaller channel. This is due to the fact that the invasion threshold capillary pressure, i.e. the pressure difference between the gas and the liquid that must be reached for the meniscus to recede in a channel, is inversely proportional to the channel size (according to Young–Laplace law, $P_{cth} = P_g - P_\ell = \dfrac{4\gamma\cos\theta}{d}$, where γ is the surface tension, θ the contact angle and d the channel equivalent diameter). Evaporation takes place not only in the channel where the meniscus is receding but also from the immobile meniscus, which means that a liquid flow directed toward the immobile meniscus occurs between the receding meniscus and the immobile meniscus. This liquid flow exactly balances the evaporation rate from the immobile meniscus. Hence, the liquid is pumped by capillarity toward the immobile meniscus. Generalization of this process to liquid clusters actually corresponds to the invasion percolation rule stating that the meniscus to recede in a cluster in the next invasion step is the one located at

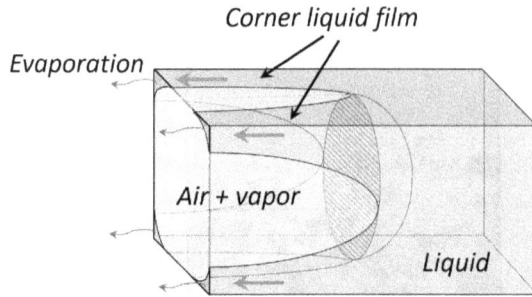

FIGURE 1.6 Illustration of corner liquid films in a capillary tube of polygonal cross section. The liquid transport in the corner films greatly enhances evaporation compared to the situation where the corner films cannot develop (as in a circular tube, for example).

the entrance of the largest interfacial pore (an interfacial pore is a pore with a meniscus at the boundary of the cluster).

In addition, the analysis of the experiments illustrated in Figure 1.4 indicated that the observed evaporation rate could not be explained considering only vapor diffusion in the pores invaded by the gas phase (Laurindo and Prat, 1998). This led to identify the flow in the liquid films present in the corners of the gas channels (Figure 1.6) as an important transport mechanism in the considered system. The impact of the corner films was later very clearly evidenced from experiments with a single tube of square cross-section (Chauvet et al., 2009). More information on this topic can be found in Chapter 2 of the book where it is shown that the corner film greatly enhances the evaporation in a square tube compared to a cylindrical tube of similar cross-section surface area.

This type of liquid films can be referred to as secondary capillary structures (SCS) since their presence is due to capillary effects and these are more difficult to visualize than the liquid clusters, which therefore can be referred to as the primary capillary structures. In fact, several types of SCS can be identified since they depend on the local geometry of the pore space and the latter can vary significantly from one porous medium to the other. Among the other SCS which have been visualized and studied in drying, one can mention the capillary bridges (Chen et al., 2017, 2018; Cejas et al., 2017) and the capillary rings (Vorhauer et al., 2015; Kharaghani et al., 2021). SCS are illustrated in Figure 1.7 and further discussed in Section 1.4 in relation to the modeling.

Experiments in microfluidic devices have also been performed to study the impact of correlations in the pore size distribution. The case of a dual-porosity channel networks is considered in Pillai et al. (2009). This first allows illustrating the mechanism of preferential invasion in the coarse porous medium (Pillai et al., 2009; Diouf et al., 2018 and references therein), which can be seen as a macroscopic equivalent of the preferential invasion in the largest tube in the interconnected channel system in Figure 1.5. As exemplified in Figure 1.8, the textural contrast, if sufficiently strong, can lead to instabilities in the invasion process. In the conventional picture of drying, a pore that has been invaded by the gas phase is not re-invaded later by the liquid phase during the drying process. However, in the case of the system illustrated in

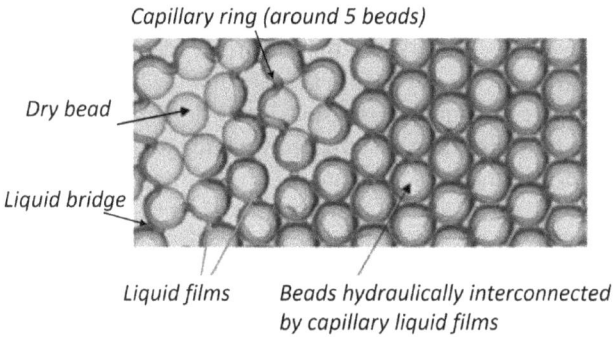

FIGURE 1.7 Secondary capillary structures (SCS) formation during drying of a model porous medium made of glass beads with liquid films, liquid bridges and capillary rings (IMFT, unpublished).

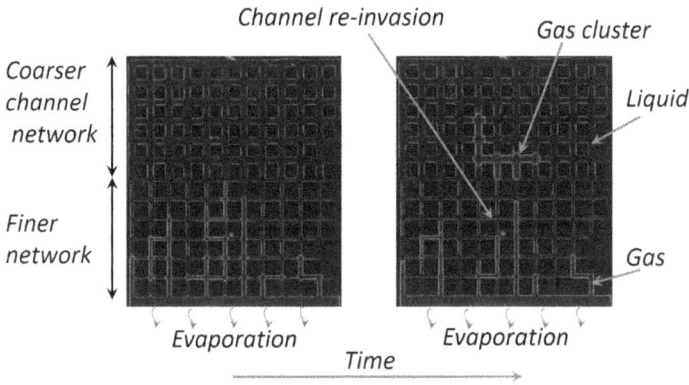

FIGURE 1.8 Capillary instability (snap-off) during the evaporation in a dual-porosity interconnected channel square network. A total of 11 channels are re-invaded by the liquid in the finer network as a result of the gas burst occurring when the gas phase reaches the coarser network from the finer network. (Image courtesy of M. Marcoux, IMFT.)

Figure 1.8, it is impossible to find a stable position of the gas–liquid interface with continuity of the gas phase between the coarse and fine networks due to the separate ranges of possible meniscus curvatures between the coarse and fine networks. As a result, the preferential invasion of the coarse medium occurs through a series of burst invasion. Each time the gas phase reaches the coarse network from the fine network, the pressure must suddenly increase in the liquid in the fine network due to the lower invasion capillary pressure thresholds in the large network. As a result of the liquid pressure difference between the coarse and fine networks, liquid flows from the coarse network to the fine network. This leads to gas invasion in the coarse network with re-invasion by the liquid of some channels in the fine network. This rapid event, analogous to a Haines jump, leads to the rupture of the gas phase continuity between the coarse and the fine network via a snap-off mechanism. Thus, in this dual-porosity system, the invasion of the coarse network by the gas phase occurs through a series of bursts into the coarse network and repeated reinvasions of the fine networks via the same invasion path. One can refer to Chapter 3 of the book for other examples of capillary instabilities and liquid re-invasion during drying.

Other aspects have also been investigated from microfluidic experiments such as the impact of thermal gradients (Vorhauer et al., 2018) and the presence of salt (e.g. Liefferink et al., 2018; Rufai and Crawshaw, 2017). One can also refer to Fantinel et al. (2017) and Biswas et al. (2018) for other studies of drying with microfluidic devices and to Xu et al. (2008) for a detailed experimental investigation of the gas invasion mechanisms in a model porous medium resulting from evaporation.

1.4 MODELING

As for the other transport phenomena in porous media, the conventional approach for the modeling of evaporation in porous media is the continuum approach. In this approach, the porous medium is seen as a fictitious continuum medium and the corresponding models rely on classical concepts such as the concept of local capillary equilibrium (macroscopically represented by the capillary pressure curve also referred to as the retention curve), generalized Darcy's law (and the concepts of permeability and relative permeabilities) and effective transport coefficients (such as the effective diffusion coefficient). Drying models developed within the frame-work of the continuum approach are discussed in detail in Chapter 6 of the book and therefore not discussed further here where the emphasis is on the other model-ing approaches. As discussed in Chapter 6, a classical "validation" scheme of the continuum models is to proceed via comparisons with macroscopic drying experi-ments, often glass bead or sand bed experiments in a configuration similar to the ones depicted in Figure 1.1, see Chapter 6 and references therein. The objective of the models is to "predict" the evolution of the saturation profiles (Figure 1.3) and the evo-lution of the drying rate (Figure 1.2). As discussed in Chapter 6, this goal has actually never been truly reached since the comparisons without adjustable parameters do not lead to satisfactory results, in general. Most of the time, either the evaporation rate in the experiment is used as boundary condition (and therefore not "predicted" by the model) or parameters in the formulation of the boundary condition at the surface(s) in contact with the external air are used as adjustable parameters. In other words, continuum models can be considered as adjustable models but not as truly predic-tive models. This unsatisfactory situation explains in part why alternative models have been developed. Trying to improve the continuum models by comparison with presumably more satisfactory models is, however, not the only motivation since the models in question aim at modeling the drying process at the pore scale or directly within the pore space. The motivation was also to gain a much better understanding of the evaporation/drying process at the pore scale. One can then distinguish two main types of pore scale models: the pore network models (PNMs) and the direct simulations. Although a method based on the Volume of Fluid (VOF) approach has also been developed (e.g. Kohout and Stepanek, 2006), the latter are mostly based on the Lattice Boltzmann methods (LBM), see Chapter 4 of this book. One can also refer to Qin et al. (2019a, b) for further applications of the method. This type of approach

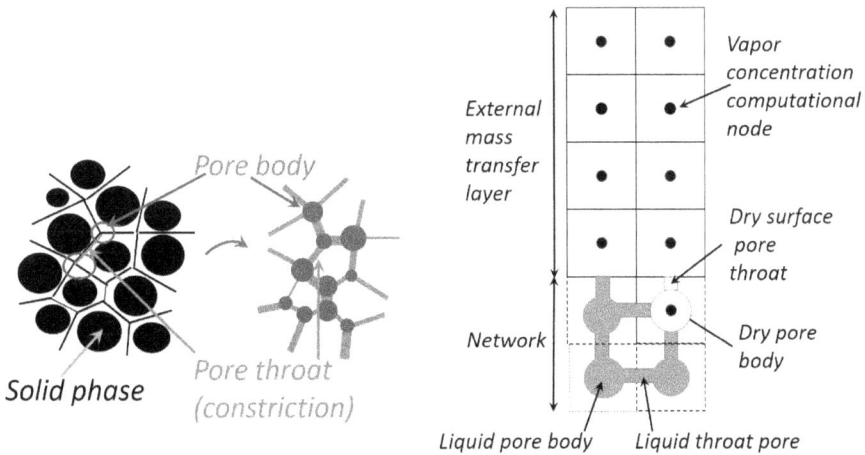

Pore body

Pore throat
(constriction)

Solid phase

Vapor concentration computational node

External mass transfer layer

Dry surface pore throat

Dry pore body

Network

Liquid pore body   Liquid throat pore

FIGURE 1.9 Left: Schematic of pore network approach. Right: Schematic of square network with vapor concentration computational nodes in external mass transfer layer.

allows a detailed simulation of the gas–liquid interface evolution within the pore space at the pore scale during evaporation. However, computational times are high, and for this reason, essentially two-dimensional systems have been studied so far with this method, see however (Amin Safi et al., 2017) for an exemple of simulations in a 3D system. Nevertheless, it can be expected that such direct approaches and the associated outcomes will gain in importance in the next future. Compared to direct simulations, PNMs allow performing simulations for a much less computational cost and have become a relatively popular investigation tool. As sketched in Figure 1.9, a PNM is based on a simplified representation of the pore space as a network of pore bodies connected by narrower pores or passages, referred to as pore throats.

In principle, the pore network can be extracted from a digital image of the microstructure of the porous medium of interest obtained, for instance, by X-ray tomography (e.g. Dong and Blunt, 2009 and references therein). However, obtaining the pore network from a 3D digital image of a porous medium microstructure is not necessarily an easy task depending on the considered porous medium. When the goal is to get insights into a given process rather than a given porous medium, simplified pore networks can be considered, such as square networks in 2D or cubic networks in 3D. In these simplified pore networks, also referred to as structured networks, the pore bodies are set at the nodes of a regular square mesh (2D) or cubic mesh (3D), whereas the pore throats are channels or tubes between two adjacent pore bodies (Figure 1.9). The essential ingredient which is kept in the structured networks is the fact that the throat sizes vary from one throat to the other. Most often, the throat sizes are simply randomly distributed according to a given probability law. Similarly, the pore body size also varies from one pore body to the other. Most of the drying simulations performed on pore networks so far have been performed considering structured pore networks (see the reviews in Prat, 2002, 2011; Metzger, 2019). The simplest drying algorithm on a network (Prat, 1993) combines the invasion percolation rule stating that the next throat to be invaded in a liquid cluster

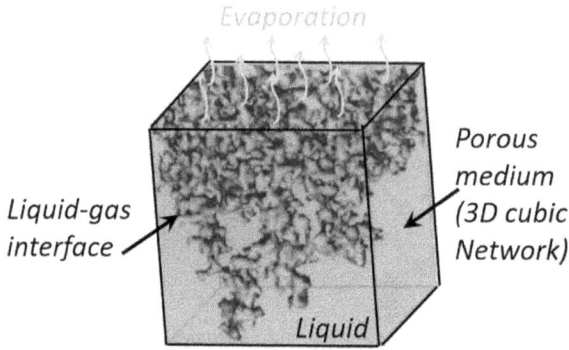

FIGURE 1.10   Illustration of the gas–liquid distribution at the beginning of the CRP in the capillary regime computed in a 3D cubic network. (Adapted from Le Bray and Prat (1999).)

subject to evaporation is the cluster interfacial pore of largest size (an interfacial pore is a pore with a meniscus) with the evaluation of the evaporation rate from each liquid cluster deduced from the computation of the vapor concentration at each pore body occupied by the gas phase in the network. To deal with the coupling with the external mass transfer, two options have been considered. The simplest one is to express the mass transfer rate from the pores at the network surface as a function of the difference in the vapor concentration at the pore and in the surrounding air and a local mass transfer coefficient (or equivalent) (e.g. Le Bray and Prat 1999). A somewhat more satisfactory approach is to add computational nodes in the external mass transfer domain (e.g. Laurindo and Prat, 1998; Attari Moghaddam et al., 2017b among others), as sketched in Figure 1.9.

The 1993 algorithm allows simulating the capillary regime, or more precisely, the "asymptotic" capillary regime corresponding to an extremely small capillary number since the impact of viscous effects on the fluid distribution is not considered. The obtained distribution in a cubic network is illustrated in Figure 1.10 (Le Bray and Prat 1999). As illustrated in Figure 1.10, the phase distribution is characterized by capillary fingers spanning the whole network in the absence of significant viscous or gravity effects. As reviewed in Prat (2002, 2011) (see also Metzger, 2019), the drying PNMs have been gradually improved over the years so as to include the impact of viscous forces in the liquid, gravity forces and thermal effects. Evaporation of multicomponent liquids in porous media has also been considered within the framework of PNMs (de Freitas and Prat, 2000). The case where the liquid is non-wetting has also been studied with PNM (e.g. Chapuis and Prat, 2007; Chraibi et al., 2009; Wu et al., 2014). Impact of mechanical effects (fracturing) has also been studied via PNMs (e.g. Kharaghani et al., 2011; Borgman and Holtzman, 2020).

As mentioned in Section 1.3, a major issue in drying is the possible impact of SCS (Figures 1.6 and 1.7) (e.g. Chauvet et al. 2009; Yiotis et al., 2012). In this respect, a major breakthrough was the consideration of corner films in drying PNM by Yiotis et al. (2004) (see also Prat, 2007; Lal et al., 2019) and Chapter 5 of the present book. Later, capillary rings were also incorporated in the pore network modeling of drying

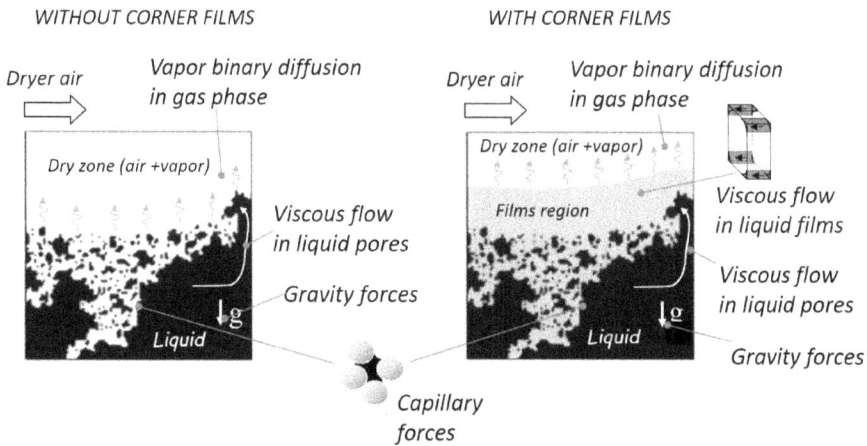

FIGURE 1.11 Main mechanisms controlling the isothermal drying of a capillary porous medium considered in the pore network models (PNMs) and schematic of the film effects. The light blue arrows indicate the position of the evaporation front in the porous medium. For a comparable overall saturation, the evaporation front is closer to the top surface of porous medium when the film effect is noticeable. As a result, the evaporation rate is greater in the presence of the corner films.

(Vorhauer et al., 2015; Kharaghani et al., 2021) as well as other capillary effects such as the so-called capillary valve effect, see Chapter 5 of the book and references therein. The effect of the corner films according to the drying PNM with corner films is sketched in Figure 1.11 where the situations with films and without films are compared for approximately the same overall saturation. As sketched in Figure 1.11, the evaporation front (blue arrows in Figure 1.11) is much closer to the open surface of porous media when the film effect is significant. Since the evaporation rate is roughly inversely proportional to the distance between the evaporation front and the open surface of the porous medium, the evaporation rate is significantly greater when the films are well developed. Also the films can be attached to the porous medium surface long after the liquid bulk connectivity to the surface has ceased. The result is a CRP much longer than predicted from the consideration of the bulk hydraulic connectivity only. Figure 1.11 also summarizes the main transport mechanisms and effects considered in the quasi-isothermal drying PNM.

In fact, various research directions can be distinguished as regards the development and use of drying PNM. One direction is to analyze the physics of drying via PNM simulations (e.g. Le Bray and Prat 1999; Yiotis et al., 2006, 2007; Wu and Zhao, 2021). A related direction is to gradually incorporate in modeling all the relevant phenomena (Figure 1.11) and to discuss the drying PNM simulations in relation to classical drying experimental results (existence of three main periods, structure of saturation profiles, etc.) (e.g. Le Bray and Prat 1999; Yiotis et al., 2006, 2007). Another direction is to explore the impact of the pore network structure on drying (e.g. Metzger, 2019 and references therein). In more recent years, drying PNM simulations have been also performed in relation to the continuum models. PNM simulations can be performed not only for computing continuum model parameters but also for exploring the consistency between PNM and continuum models with the objective of improving the continuum models (e.g. Attari-Moghaddam et al., 2017a, b; Ahmad et al., 2020; Talbi and Prat, 2022, see also Chapter 7 of the present book). Since the pore networks considered in these works are generally too small to really fulfill the criterion of length scale separation, these works also give insights into the modeling of drying in thin systems via the continuum approach, e.g. Chapter 7 of this book. The coupling with the external transfers being perhaps the most delicate issue in the evaporation modeling, works have also been dedicated to this topic via PNM simulations (e.g. Attari-Moghaddam et al., 2018; Talbi and Prat, 2021).

1.5 DISCUSSION

The simple fact that the slow drying of a capillary porous medium is still an active research topic, even for the simplest situation of the evaporation of a pure liquid, shows that the drying theory is still not completely mature. As discussed in some detail in Chapter 6 of the book, perhaps the main issue regarding the continuum models lies in the boundary or coupling conditions at the surfaces in contact with the surrounding air. At the moment, there is no proposal for the formulation of the coupling conditions based on a solid theoretical approach. In fact, the used conditions are essentially empirical and involve adjustable parameters. As a result, the drying continuum models cannot be considered as truly predictive. To improve this situation, it makes sense to try to rely on pore scale models or simulations, such as PNM or LBM. To this end, however, LBM simulations should be performed over much larger domains than the ones considered so far and in 3D. This represents a considerable computational challenge. In this respect, PNMs are more attractive due to their much less computational cost. However, the two-phase flow detailed physics can be expected to be well reproduced by the LBM simulations whereas the situation is more complex with PNM in this regard. Although it is possible to get insights into the continuum approach from PNM simulations (e.g. Attari-Moghaddam et al., 2017a), a difficulty is that the PNMs have never been quantitatively compared with classical drying experiments, such as the archetypical experiments with glass bead packing or sand beds. Such a comparison would notably allow getting insights on the likely role of the SCS (capillary bridges, rings, corner films, etc.) in these systems. To this end, at least four challenges must be met. The first one is to develop unstructured drying pore networks based on real microstructure digital images (e.g. Wang et al., 2012). The second one is to improve the computation of the external mass transfer. As depicted in Figure 1.9, the distance between computational nodes in the external layer is typically equal to the mean distance between the pores in the network in the drying PNM studies developed so far. This is sufficient for qualitative results but not if the goal is quantitative comparisons with experimental results. To this end, much more refined computational grids should be considered in the external layer (e.g. Weishaupt et al., 2019), at least in the vicinity of the porous medium surface. The third one is the adequate modeling of SCS. Although quite interesting, the drying PNM with SCS proposed by Yiotis et al. (2004) is based on the simplified assumption of well-connected corner films throughout the film region. As discussed in Chapters 3 and 5 of the book, the real situation can be significantly more complex. Modeling the SCS during drying in a random packing of particles is still a challenge. The fourth one is the computational performance. So far, PNM simulations have been performed over relatively small networks (the largest one had only $80 \times 80 \times 80$ pore bodies (Yiotis et al., 2006), whereas substantially smaller networks were considered in many studies). To perform comparison with classical experiments and meet the requirement of length scale separation, substantially larger networks should be considered. This implies to develop high performance computing versions of the PNM codes.

Although not addressed in the present book, another interesting research direction lies in the consideration of drying situations where particles (e.g. Keita et al., 2013; Qin et al., 2019b), polymers (e.g. Faiyas et al., 2017) or one or more dissolved salts (e.g. Eloukabi et al., 2013; Talbi and Prat, 2022) are present in the solution. This type of situation can lead to a rather complex phenomenology (e.g. Licsandru et al., 2019, 2022), with many open questions. Here also, experiments in single channels (Naillon et al., 2015), microfluidic devices or micromodels (e.g. Sghaier et al., 2014; Liefferink et al., 2018; Rufai and Crawshaw, 2017), model porous media (e.g. Keita et al., 2013; Nachshon et al., 2011; Bergstad et al., 2017), continuum models (e.g. Lazhar et al., 2020), LBM simulations (e.g. Qin et al., 2019b) and PNM simulations (e.g. Dashtian et al., 2018; Ahmad et al., 2021) are helpful to improve our understanding of these situations and our prediction capabilities.

The above applies to capillary porous media. However, evaporation with salt crust formation, for example, indicates that the salt crust is porous and can contain submicronic pores (Licsandru et al., 2022). In other words, considering only capillary porous media is not sufficient. The hygroscopic porous media should be addressed, not only, of course, in relation to the evaporation of saline solutions or of solutions containing particles but because they correspond to a very important category of porous media, including notably wood, concrete or tight rocks. Slow drying of hygroscopic porous media shares common features with drying of capillary porous media but also presents specific differences, notably due to the significance of adsorption phenomena (and the possible transport in the bound liquid) and Kelvin effect. Also, the physics of drying can be markedly different when the porous medium structure is in such a way that cavitation effects occur (e.g. Vincent et al., 2014). Although a drying PNM considering the Kelvin effect has been developed (Maalal et al., 2021), developing PNM for hygroscopic porous media is a widely open topic. So far, essentially continuum models have been developed (not considering here the molecular dynamics (MD) simulations as they usually consider the extremely small spatial and time domains) in conjunction with macroscopic experiments. Studies of evaporation in nanofluidic systems are also scarce (e.g. Eijkel et al., 2005; Duan et al., 2012). In other words, hygroscopic porous media would deserve an attention and efforts equivalent to the ones devoted to the study of capillary porous media over the last three decades. However, the study of hygroscopic porous media can be considered as significantly more difficult because of the possible significance of nanoscale phenomena that can be neglected in capillary porous media and also, from a practical standpoint, because the observation of the phase distribution during the drying process, for instance, is much more challenging in pores in the sub-micronic range.

1.6 SUMMARY

Evaporation in capillary porous media has been studied over the years using a variety of experimental and numerical tools. Up to the eighties, the studies were mainly based on continuum models and macroscopic experiments. From the nineties, experiments with microfluidic systems and PNM simulations have led to a significantly greater understanding of the evaporation process at the pore scale and at the pore network scale. More recently, LBM simulations and the easier access to digital images of porous microstructures via modern imagery techniques have expanded the research tools and offer new prospects.

Nevertheless, predicting the drying or evaporation process in a porous medium remains a challenge, even for the somewhat simpler porous media considered in this chapter referred to as the capillary porous media. This is because the continuum approach to drying remains incomplete, mainly due to the lack of fully predictive and rigorous theory of the transfers in the interfacial region with the surrounding air. So far, the numerous insights gained from the microfluidic experiments and the pore scale simulations, based on PNM or direct simulations (LBM), have not yet had a substantial impact regarding the formulation of the coupling with the external transfer within the framework of the continuum approach. However, one cannot directly rely on PNM or LBM simulations in most applications due to the much too high computational costs. Thus, there is still a need of improving the continuum models to deal with many situations encountered in practice. However, particular systems, such as the thin systems characterized by a limited number of pores over their thickness, are good candidates for direct studies by means of PNM or LBM simulations.

The current theory of drying also faces a challenge with the SCS, such as the capillary bridges, corner films or capillary rings that form in the regions invaded in the bulk by the gas phase. There is a need for more accurate modeling of the SCS in the PNM and LBM simulations, whereas the explicit incorporation of the SCS in continuum models is still an open question.

In relation to the state of the art presented in this chapter, the next chapters in the book will present modern illustrations of the interest of single tube or microfluidic experiments (Chapters 2 and 3), LBM simulations (Chapter 4), PNM simulations (Chapters 5 and 7) and continuum models (Chapter 6).

ACKNOWLEDGMENTS

Discussions and collaborative research with many students and colleagues from various countries on evaporation/drying in porous media over the years have directly or indirectly greatly contributed to this review. They are warmly thanked.

REFERENCES

Adamson, A.W. 1990. *Physical Chemistry of Surfaces*, 5th ed., Wiley, New York.
Ahmad, F., Talbi, M., Prat, M., Tsotsas, E., Kharaghani, A. 2020. Non-local equilibrium continuum modeling of partially saturated drying porous media: Comparison with pore network simulations, *Chem. Eng. Sci.* 228: 115957.

Ahmad, F.,Rahimi, A., Tsotsas, E., Prat, M., Kharaghani, A; 2021. From micro-scale to macro-scale modeling of solute transport in drying capillary porous media, *Int. J. of Heat and Mass Transfer* 165: 120722.

Amin Safi, M., Prasianakis, N.I., Mantzaras, J., Lamibrac A., Büchi, F.N. 2017. Experimental and pore-level numerical investigation of water evaporation in gas diffusion layers of polymer electrolyte fuel cells, *Int. J. Heat Mass Transf.* 115: 238–249.

Attari-Moghaddam, A., Kharaghani, A., Tsotsas, E., Prat, M. 2017a. Kinematics in a slowly drying porous medium: Reconciliation of pore network simulations and continuum modeling, *Phys. Fluids* 29(2): 022102.

Attari-Moghaddam, A., Kharaghani, A., Tsotsas, E., Prat, M. 2018. A pore network study of evaporation from the surface of a drying non-hygroscopic porous medium, *AIChE J.* 64(4), 1435–1447.

Attari-Moghaddam, A., Prat, M., Tsotsas, E., Kharaghani, A. 2017b. Evaporation in capillary porous media at the perfect piston-like invasion limit: Evidence of non-local equilibrium effects, *Water Resour. Res.* 53(12): 10433–10449.

Benavente, D., Sanchez-Moral, S., Fernandez-Cortes, A., Cañaveras, J. C., Elez, J., Saiz-Jimenez, C. 2011. Salt damage and microclimate in the Postumius Tomb, Roman Necropolis of Carmona, Spain. *Environ. Earth Sci.* 63: 1529–1543.

Bergstad, M., Or, D., Withers, P.J., Shokri, N. 2017. The influence of NaCl concentration on salt precipitation in heterogeneous porous media, *Water Resour. Res.* 53(2): 1702–1712.

Biswas, S., Fantinel, P., Borgman, O., Holtzman, R., Goehring, L. 2018. Drying and percolation in spatially correlated porous media. *Phys. Rev. Fluids* 3: 124307.

Borgman, O., Holtzman, R. 2020. Impact of matrix deformations on drying of granular materials. *Int. J. Heat Mass Transf.* 153: 119634.

Brutsaert, W., Chen, D. 1995. Desorption and the two stages of drying of natural tallgrass prairie, *Water Resour. Res.* 31(5): 1305–1313.

Cejas, C.M., Castaing, J.C., Hough, L., Frétigny, C., Dreyfus, R. 2017. Experimental investigation of water distribution in a two-phase zone during gravity-dominated evaporation. *Phys. Rev. E* 96: 062908.

Chapuis, O., Prat, M. 2007. Influence of wettability conditions on slow evaporation in two-dimensional porous media. *Phys. Rev. E* 75: 046311.

Chauvet, F., Duru, P., Geoffroy, S., Prat, M. 2009. Three periods of drying of a single square capillary tube. *Phys. Rev. Lett.* 103: 124502.

Chen, C., Duru, P., Joseph, P., Geoffroy, S., Prat, M. 2017. Control of evaporation by geometry in capillary structures. From confined pillar arrays in a gap radial gradient to phyllotaxy-inspired geometry. *Sci. Rep.* 7(1): 15110.

Chen, C., Joseph, P., Geoffroy, S., Prat, M., Duru, P. 2018. Evaporation with formation of liquid bridges chains. *J. Fluid Mech.* 837: 703–728.

Chraibi, H., Prat, M., Chapuis, O. 2009. Influence of contact angle on slow evaporation in two dimensional porous media. *Phys. Rev. E* 79: 026313.

Coussot, P. 2000, Scaling approach of the convective drying of a porous medium. *Eur. Phys. J. B* 15: 557–566.

Dashtian, H., Shokri, N. Sahimi. M. 2018. Pore-network model of evaporation-induced salt precipitation in porous media: The effect of correlations and heterogeneity. *Adv. Water Resour.* 112: 59–71.

de Freitas, D.S., Prat, M. 2000. Pore network simulation of evaporation of a binary liquid from a capillary porous medium. *Transp. Porous Med.* 40: 1–25.

Diouf, B., Geoffroy, S., Abou Chakra, A., Prat, M. 2018. Locus of first crystals on the evaporative surface of a vertically textured porous medium. *Eur. Phys. J. Appl. Phys.* 81(1): 11102.

Dong, H., Blunt, M.J. 2009. Pore-network extraction from micro-computerized-tomography images. *Phys. Rev. E* 80: 036307.

Duan, C., Karnik, R., Lu, M.C., Majumdar, A. 2012. Evaporation-induced cavitation in nano-fluidic channels. *Proc. Natl. Acad. Sci.* 109(10): 3688–3693.

Eijkel, J. C. T., Dan, B., Reemeijer, H.W., Hermes, D.C, Bomer, J.G., van den Berg, A. 2005. Strongly accelerated and humidity-independent drying of nanochannels induced by sharp corners. *Phys. Rev. Lett.* 95: 256107.

Eloukabi, H., Sghaier, N., Ben Nasrallah, S., Prat, M. 2013. Experimental study of the effect of sodium chloride on drying of porous media: The crusty-patchy efflorescence transition, *Int. J. Heat Mass Transf.* 56: 80–93.

Engøy, T., Feder, J., Jøssang, T. 1991. Counter-flow in drying of competing pores. *Phys. Scr.* T38: 99–102.

Faiyas, A.P.A., Erich S. J. F., Huinink, H. P., Adan, O. C. G., Transport of a water-soluble polymer during drying of a model porous media, *Dry. Technol.* 35(15): 1874–1886.

Fang, A., Chen, Y., Wu, L. 2021. Modeling and numerical investigation for hygrothermal behavior of porous building envelope subjected to the wind driven rain. *Energy Build.* 231, 110572.

Fantinel, P., Borgman, O., Holtzman, R., Goehring, L. 2017. Drying in a microfluidic chip: Experiments and simulations. *Sci. Rep.* 7: 15572.

Ghiringhelli, E. Ongoing Ph.D Thesis, University of Toulouse.

Gupta, S., Huinink, H.P., Prat, M., Pel, L., Kopinga, K. 2014. Paradoxical drying due to salt crystallization. *Chem. Eng. Sci.* 109: 204–211.

Hassani, A., Azapagic, A., Shokri N. 2021. Global predictions of primary soil salinization under changing climate in the 21st century. *Nat. Commun.* 12(1), 1–17.

Keita, E., Faure, P., Rodts, S., Coussot, P. 2013. MRI evidence for a receding-front effect in drying porous media. *Phys. Rev. E* 87: 062303.

Kharaghani, A., Mahmood, H.T., Wang, Y., Tsotsas, E. 2021. Three-dimensional visualization and modeling of capillary liquid rings observed during drying of dense particle packings. *Int. J. Heat Mass Transf.* 177: 121505.

Kharaghani, A., Metzger, T., Tsotsas, E. 2011. A proposal for discrete modeling of mechanical effects during drying, combining pore networks with DEM. *AIChE J.* 57(4): 872–885.

Kohout, M., Grof, Z., Stepanek, F. 2006. Pore-scale modelling and tomographic visualization of drying in granular media. *J. Colloid Interface Sci.* 299: 342–351.

Lal, S., Prat, M., Plamondon, M., Poulikakos, L., Partl, M.N., Derome, D., Carmeliet, 2019. A cluster-based pore network model of drying with corner liquid films, with application to a macroporous material. *Int. J. Heat Mass Transf.* 140: 620–633.

Laurindo, J.B., Prat, M. 1996. Numerical and experimental network study of evaporation in capillary porous media. Phase distributions. *Chem. Eng. Sci.* 51(23): 5171–5185.

Laurindo, J.B., Prat, M. 1998. Numerical and experimental network study of evaporation in capillary porous media. Drying rates. *Chem. Eng. Sci.* 53(12): 2257–2269.

Lazhar, R., Najjari, M., Prat, M. 2020. Combined wicking and evaporation of NaCl solution with efflorescence formation: The efflorescence exclusion zone. *Phys. Fluids* 32(6): 067106.

Le Bray, Y., Prat, M. 1999. Three dimensional pore network simulation of drying in capillary porous media, *Int. J. of Heat and Mass Tr.* 42: 4207-4224.

Lehmann, P., Assouline, S., Or Dani, 2008. Characteristic lengths affecting evaporative drying of porous media. *Phys. Rev. E* 77: 056309

Lehmann, P., Or, D. 2013. Effect of wetness patchiness on evaporation dynamics from drying porous surfaces. *Water Resour. Res.* 49: 8250–8262.

Lenormand, R., Touboul, E., Zarcone, C. 1988. Numerical models and experiments on immiscible displacements in porous media, *J. Fluid Mech.* 189: 165–187.

Licsandru, G., Noiriel, C., Duru, P., Geoffroy, S., Abou-Chakra, A., Prat, M. 2019. Dissolution-precipitation-driven upward migration of a salt crust. *Phys. Rev. E* 100: 032802.

Licsandru, G., Noiriel, C., Geoffroy, S., Abou-Chakra, A., Duru, P., Prat, M. 2022. Evaporative NaCl salt crust: Pore size, permeability, detachment mechanism, reduced evaporation, under review.

Liefferink, R.W., Naillon, A., Bonn, D., Prat, M., Shahidzadeh, N. 2018. Single layer porous media with entrapped minerals for microscale studies of multiphase flow. *Lab Chip* 18(7): 1094–1104.

Maalal, O., Prat, M., Lasseux, D. 2021. Pore network model of drying with Kelvin effect, *Phys. Fluids* 33(2): 027103.

Metzger, T. 2019. A personal view on pore network models in drying technology. *Dry. Technol.* 37: 497–512.

Mujumdar, A.S. 2006. *Handbook of Industrial Drying*, 3rd Edition, CRC Press, Boca Raton.

Muller, N., Qi, R., Mackie E., Pruess, K., Blunt, M.J. 2009. CO_2 injection impairment due to halite precipitation, *Energy Procedia* 1(1): 3507–3514.

Nachshon, U., Weisbrod, N., Dragila, M.I., Grader, A. 2011. Combined evaporation and salt precipitation in homogenious and heterogenious porous media, *Water Resour. Res.* 47: W03513.

Naillon, A., Duru, P., Marcoux, M., Prat, M. 2015. Evaporation with sodium chloride crystallization in a capillary tube. *J. Cryst. Growth* 422: 52–61.

Or, D., Lehmann, P., Shahraeeni E., Shokri, N. 2013. Advances in soil evaporation physics— A review. *Vadose Zone J.* 12(4): vzj2012.0163.

Pel L., Brocken, H., Kopinga, A.1996. Determination of moisture diffusivity in porous media using moisture concentration profiles. *Int. J. Heat Mass Transf.* 39(6): 1273–1280.

Pel L., Ketelaarss, A. A. J., Adan, O.C.G., Van Well, A. A. 1993. Determination of moisture diffusivity in porous media using scanning neutron radiography. *Int. J. Heat Mass Transf.* 36(5): 1261–1267.

Pillai, K.M., Prat, M., Marcoux, M. 2009. A study on slow evaporation of liquids in a dual- porosity porous medium using square network model. *Int. J. Heat Mass Transf.* 52: 1643–1656.

Prat, M. 1993. Percolation model of drying under isothermal conditions in porous media. *Int. J. of Multiphase Flow.* 19 (4): 691-704.

Prat, M. 2002. Recent advances in pore-scale models for drying of porous media. *Chem. Eng. J.* 86: 153–164.

Prat, M. 2007. On the influence of pore shape, contact angle and film flows on drying of capillary porous media. *Int. J. Heat Mass Transf.* 50: 1455–1468.

Prat, M. 2011. Pore network models of drying, contact angle and films flows, *Chem. Eng. Technol.* 34(7): 1029–1038.

Prat, M., Bouleux, F. 1999. Drying of capillary porous media with stabilized front in two-dimensions, *Phys. Rev. E* 60: 5647–5656.

Qin, F., Del Carro, L., Moqaddam, A.M., Kang, Q., Brunschwiler, T., Derome, D., Carmeliet, J. 2019a. Study of non-isothermal liquid evaporation in synthetic micro-pore structures with hybrid lattice Boltzmann model. *J. Fluid Mech.* 866: 33–60.

Qin, F., Moqaddam, A.M., Kang, Q., Derome, D., Carmeliet, J. 2019b. LBM simulation of self-assembly of clogging structures by evaporation of colloidal suspension in 2D porous media. *Trans. Porous Med.* 128: 929–943.

Rufai, A., Crawshaw, J. 2017. Micromodel observations of evaporative drying and salt deposition in porous media. *Phys. Fluids* 29: 126603.

Schlünder, E.U. 1988. On the mechanism of the constant drying rate period and its relevance to diffusion controlled catalytic gas phase reactions, *Chem. Eng. Sci.* 43(10): 2685–2688.

Schrage, R.W. 1953. *A Theoretical Study of Interphase Mass Transfer*, Columbia University Press, New York.

Sghaier, N., Geoffroy, S., Prat, M., Eloukabi, H., Ben Nasrallah, S. 2014. Evaporation driven growth of large crystallized salt structures in a porous medium. *Phys. Rev. E* 90: 042402.

Shah, F.H., Hadim, H.A., Korfiatis G.P., 1995. Laboratory studies of air stripping of VOC-contaminated soils. *J. Soil Contam.* 4(1): 493–109.

Shaw, T.M., 1987. Drying as an immiscible displacement process with fluid counterflow, *Phys. Rev. Lett.* 59: 1671.

Shokri, N., Or D., 2011.What determines drying rates at the onset of diffusion controlled stage-2 evaporation from porous media? *Water Resour. Res.* 47: W09513.

Stauffer, D., Aharony, A. 1992. *Introduction to Percolation Theory*, Taylor & Francis, London.

Suzuki, M.; Maeda S. 1968. On the mechanism of drying of granular beds, *J. Chem. Eng. Jpn.* 1(1): 26–31.

Talbi, M., Prat, M. 2019. About Schlünder's model: A numerical study of evaporation from partially wet surfaces. *Dry. Technol.* 37(5): 513–524.

Talbi, M., Prat, M. 2021. Coupling between internal and external mass transfer during stage-1 evaporation in capillary porous media: Interfacial resistance approach, *Phys. Rev. E* 104, 055102.

Talbi, M., Prat, M. 2022. Percolating and non-percolating liquid phase continuum model of drying in capillary porous media with application to solute transport in the very low Péclet number limit. *Phys. Rev. Fluids* 7: 014306.

Thiery, J., Rodts, S., Weitz, D.A., Coussot, P. 2017. Drying regimes in homogeneous porous media from macro-to nanoscale, *Phys. Rev. Fluids* 2: 074201.

Tsimpanogiannis, I.N., Yortsos, Y.C., Poulou, S., Kanellopoulos, N., Stubos, A.K., 1999. Scaling theory of drying in porous media, *Phys. Rev. E* 59(4): 4353–4365.

Van Brakel, J. 1980. Mass Transfer in Convective Drying. *Adv. Dry.* 1: 217–267.

Vincent, O., Sessoms, D.A., Huber, E.J., Guioth, J., Stroock, A.D. 2014. Drying by cavitation and poroelastic relaxations in porous media with macroscopic pores connected by nanoscale throats. *Phys. Rev. Lett.* 113(13): 134501.

Vorhauer, N., Tsotsas, E., Prat, M. 2018. Temperature gradient induced double stabilization of the evaporation front within a drying porous medium, *Phys. Rev. Fluids* 3: 114201.

Vorhauer, N., Wang, Y., Karaghani, A., Tsotsas, E., Prat, M. 2015. Drying with formation of capillary rings in a model porous medium. *Trans. Porous Med.* 110: 197–223.

Wang, M., Dong, X., Liu, J., Zhao, X. 2022.Experimental and numerical investigation of wicking and evaporation performance of fibrous materials for evaporative cooling. *Energy Build.* 255: 111675.

Wang, Y., Kharaghani, A., Metzger, T., Tsotsas, E. 2012. Pore network drying model for particle aggregates: assessment by X-ray microtomography. *Dry. Technol.* 30: 1800–1809.

Weishaupt, K., Joekar-Niasar, V., Helmig, R. 2019. An efficient coupling of free flow and porous media flow using the pore-network modeling approach. *J. Comput. Phys.* X(1): 100011.

Whitaker, S. 1991. Role of the species momentum equation in the analysis of the Stefan diffusion tube. *Ind. Eng. Chem. Res.* 30(5): 978–983.

Wilkinson, D., Willemsen, J.F. 1983. Invasion percolation: a new form of percolation theory, *J. Phys. A: Math. Gen.* 16: 3365–3376.

Wu, R., Cui, G.M., Chen, R. 2014. Pore network study of slow evaporation in hydrophobic porous media. *Int. J. Heat Mass Transf.* 68: 310–323.

Wu, R., Zhao, C.Y. 2021. Distribution of liquid flow in a pore network during evaporation. *Phys. Rev. E* 104: 025107.

Xu, L., Davies, S., Schofield, A.B., Weitz, D.A.2008. Dynamics of drying in 3D porous media. *Phys. Rev. Lett.* 101: 094502.

Yiotis, A.G., Boudouvis, A.G., Stubos, A.K., Tsimpanogiannis, I.N., Yortsos, Y.C. 2004. Effect of liquid films on the drying of porous media, *AIChE J.* 50: 2721–2737.

Yiotis, A.G., Salin, D., Tajerand, E.S., Yortsos, Y.C. 2012. Drying in porous media with gravity-stabilized fronts: Experimental results. *Phys. Rev. E* 86: 026310.

Yiotis, A.G., Tsimpanogiannis, I.N., Stubos, A.K., Yortsos, Y.C. 2006. Pore-network study of the characteristic periods in the drying of porous materials. *J. Colloid Inter. Sci.* 297(2): 738–748.

Yiotis, A.G., Tsimpanogiannis, I.N., Stubos, A.K., Yortsos, Y.C. 2007. Coupling between external and internal mass transfer during drying of a porous medium. *Water Res. Resour.* 43(6): W06403.

Zenyuk, I.V., Lamibrac, A., Eller, J., Parkinson, D.Y., Marone, F., Büchi, F.N., Weber, A.Z. 2016. Investigating evaporation in gas diffusion layers for fuel cells with x-ray computed tomography. *J. Phys. Chem. C* 120(50): 28701–28711.

2 Evaporation from Straight Capillary Tubes

Rui Wu
Shanghai Jiao Tong University

Evangelos Tsotsas
Otto-von-Guericke-Universitat Magdeburg

CONTENTS

2.1 INTRODUCTION

Although much progress, both experimentally and numerically, has been made on the evaporation of porous media, the accurate prediction of the evaporation rate is still a challenge (Chauvet et al., 2009), mainly owing to the complex pore structures and complicated mass transport processes in the pores. The void space of a porous material can be considered as a collection of pores of various structures and sizes. As a result, disclosing the evaporation of a single pore is an important first step for understanding the evaporation of porous media composed of interconnected pores.

During evaporation of a porous material, the initially liquid-filled pores are gradually occupied by gas. In the course of gas invasion into pores in the porous medium, liquid can be retained in the pore corners. Whether the liquid corner films can form or not depends on the shape of the corner and the wettability of the wall surface. For instance, the liquid corner films cannot form in the pores with hydrophobic surfaces. To understand the evaporation of pores with and without liquid corner films, two types of straight tubes are considered in this chapter. The first one is of circular cross section for which the liquid corner films are absent. The second one is a tube of square cross section for which the liquid corner films can be formed if the contact angle θ is inferior to a critical contact angle $\theta_c = \dfrac{\pi}{2} - \alpha = 45°$ (Concus and Finn, 1969). Here, α is the corner half-angle. In the case of a square tube, $\alpha = \dfrac{\pi}{4}$.

DOI: 10.1201/9781003011811-2

In what follows, the evaporation of a straight tube of circular cross section is considered. In Section 2.3, the evaporation of a straight tube of square cross section is presented. After the introduction of evaporation of a single tube, evaporation of a bundle of tubes is illustrated in Section 2.4. Finally, the conclusion is drawn in Section 2.5.

2.2 EVAPORATION OF A STRAIGHT TUBE OF CIRCULAR CROSS SECTION

The mass transfer driven evaporation of a straight tube of circular cross section is illustrated in Figure 2.1 (Bird, 1960; Whitaker, 1991). During evaporation, liquid in the tube vaporizes and diffuses into stagnant air. The dissolution of air in liquid is negligible. Based on the vapor mass balance equation, the transfer of vapor in the tube can be described as

$$\nabla \cdot N_v + \frac{\partial c_v}{\partial t} = 0 \tag{2.1}$$

where N_v, c_v, and t are the vapor molar flux, vapor molar concentration, and time, respectively. The vapor molar flux N_v is equal to $c_v v_v$, where v_v is the vapor velocity.

Considering the vapor diffusion as quasi-steady and the diffusion in the tube as a one-dimensional process along the z-direction, Eq. (2.1) is reduced to

$$\frac{dN_{v,z}}{dz} = 0 \tag{2.2}$$

Similarly, the diffusion of air can be described as

$$\frac{dN_{a,z}}{dz} = 0 \tag{2.3}$$

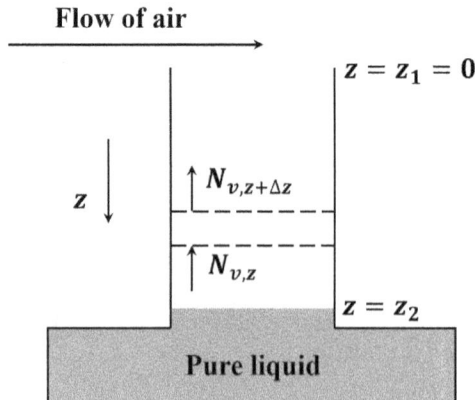

FIGURE 2.1   Schematic of evaporation in a straight circular tube.

Equations (2.2) and (2.3) indicate that N_v and N_a are constant along the z-direction. Since air has negligible solubility in liquid, air flux at the gas–liquid interface, i.e., $z = z_2$, is zero. Thus, $N_{a,z} = 0$.

The diffusion rate of vapor along the z-direction is

$$J_{v,z} = c_v \left(v_{v,z} - V_z \right) = -D \frac{dc_v}{dz} \tag{2.4}$$

where $v_{v,z}$ and V_z are the velocity of vapor and the molar averaged velocity in the z-direction, respectively. D is the vapor diffusivity. The molar averaged velocity V_z is determined as

$$V_z = \frac{c_v v_{v,z} + c_a v_{a,z}}{c_v + c_a} = y_v v_{v,z} + y_a v_{a,z} \tag{2.5}$$

where $y_v = c_v/c_g$ and $y_a = c_a/c_g$ are molar fractions of vapor and air, respectively. Here, $c_g = c_v + c_a$ is the molar concentration of the gas mixtures of vapor and air. The gas mixtures can be considered as the ideal gas. Hence, the molar concentration c_g can be expressed as

$$c_g = \frac{P_g}{R_u T} \tag{2.6}$$

where P_g is the pressure of the gas mixture, T the temperature, and R_u the universal gas constant.

Based on the definition of $N_{v,z} = c_v v_{v,z}$, we can get

$$N_{v,z} = -D \frac{dc_v}{dz} + c_v V_z = -c_g D \frac{dy_v}{dz} + y_v \left(N_{v,z} + N_{a,z} \right) \tag{2.7}$$

Since $N_{a,z} = 0$, Eq. (2.7) is reduced to

$$N_{v,z} = -c_g D \frac{dy_v}{dz} + y_v N_{v,z} \tag{2.8}$$

Integrating Eq. (2.8) from the inlet of tube ($z = z_1$) to the gas–liquid interface ($z = z_2$) yields

$$N_{v,z} \int_{z_1}^{z_2} dz = c_g D \int_{y_{v,1}}^{y_{v,2}} -\frac{dy_v}{1 - y_v} \tag{2.9}$$

for which we get

$$N_{v,z} = -\frac{c_g D}{z_2 - z_1} \ln \frac{\left(1 - y_{v,2}\right)}{\left(1 - y_{v,1}\right)} \tag{2.10}$$

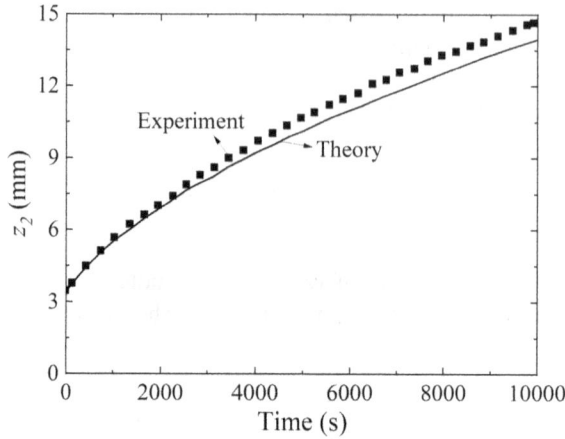

FIGURE 2.2 Comparison of the variations of the bulk meniscus position, i.e., z_2, illustrated in Figure 2.1, with time obtained from experiment and Eq. (2.13).

Based on Eq. (2.10), the evaporation flux E from the tube can be obtained. If we take $z_1 = 0$, then

$$E = \rho_l \frac{dz_2}{dt} = N_{v,z} M_v = -\frac{P_g D M_v}{z_2 R_u T} \ln \frac{\left(1 - y_{v,2}\right)}{\left(1 - y_{v,1}\right)} \tag{2.11}$$

where ρ_l is the liquid density. By integrating Eq. (2.11), we can get the variation of the location of the gas–liquid interface z_2 with the time:

$$\int_0^{z_2} z_2 dz_2 = -\frac{P_g D M_v}{z_2 R_u T \rho_l} \ln \frac{\left(1 - y_{v,2}\right)}{\left(1 - y_{v,1}\right)} \int_0^t dt \tag{2.12}$$

This yields

$$z_2 = \sqrt{\frac{2 P_g D M_v}{R_u T \rho_l} \ln \frac{\left(1 - y_{v,1}\right)}{\left(1 - y_{v,2}\right)} t} \tag{2.13}$$

The prediction of Eq. (2.13) is compared against the experimental results obtained from drying of a circular tube filled with hexane (Chauvet et al., 2010). Here, $D = 0.81 \times 10^{-5} \, \text{m}^2/\text{s}^{-1}$, $M_v = 0.086 \, \text{kg/mol}^{-1}$, $\rho_l = 660.6 \, \text{kg/m}^{-3}$, $T = 296 \, \text{K}$, $P_g = 1.01 \times 10^{-5} \, \text{Pa}$, $y_{v,1} = 0$, and $y_{v,2} = 0.185$. As shown in Figure 2.2, the predicted results agree well with the experimental data.

2.3 EVAPORATION OF A STRAIGHT TUBE OF SQUARE CROSS SECTION

In the foregoing section, we present the evaporation of a straight tube of circular cross section for which the corner liquid films are absent. In this section, a theoretical

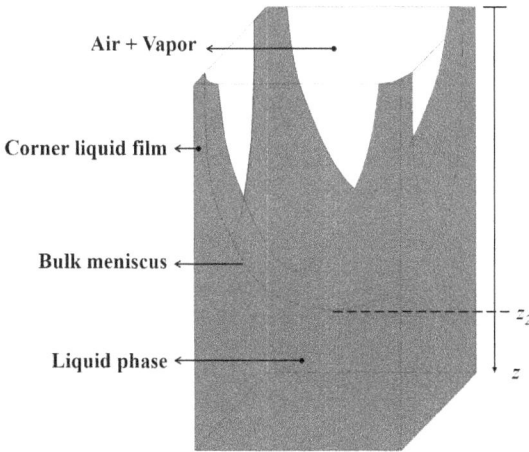

FIGURE 2.3 Drying of a capillary tube of square cross section.

model to describe the evaporation in a straight tube with corner liquid films is developed (Chauvet et al., 2010). To validate the theoretical model, evaporation experiments are performed with a straight tube of square cross section, as illustrated in Figure 2.3.

2.3.1 EXPERIMENT

The experimental setup consists of a capillary tube, held vertically and glued at one of its ends by an epoxy resin directly to a syringe tip, with the other end being opened and placed in stagnant dry air. The tube filling is controlled by a precision syringe pump (PHD 2000, Harvard Apparatus).

Two types of capillary glass tubes are used in the experiments. The first types are 10 cm long square capillary tubes made of borosilicate glass and supplied by Vitrocom. The second types of tubes are 10 cm long square capillary tubes made of borosilicate and supplied by Hilgenberg. The internal side length and wall thickness of the tubes are given in Table 2.1. As can be noticed by imaging tubes' cross sections with an optical microscope, the tubes' internal corners are not sharp. The degree of roundedness, r_0, as defined in Figure 2.4a, is estimated by fitting the internal corner shapes by quarters of circular arcs for several tube cross sections, Figure 2.4b,c. The values of r_0 for the capillary tubes used are given in Table 2.1. Note that the internal corner cross-section shapes of the Hilgenberg capillary tubes do not consist of a circular arc, contrary to the Vitrocom tubes, probably due to different fabrication processes. The roundedness for the Hilgenberg capillary tubes is smaller than that for the Vitrocom tubes. Three perfectly wetting fluids are used: Heptane, 2-propanol, and a nonvolatile silicon oil (47V5). The fluid properties are presented in Table 2.2.

The measurements rely on video imaging of the capillary tube, using an ombroscopy technique in order to detect easily the location of the interface between the liquid and the gas. Two CCD cameras (Sensicam, PCO) are used. One camera is facing

TABLE 2.1
Capillary Tubes' Characteristics

	Tube 1	Tube 2	Tube 3	Tube 4
Supplier	Vitrocom	Vitrocom	Hilgenberg	Hilgenberg
Internal dimension (mm)	1	0.4	0.95	0.49
Tube wall thickness (μm)	200	200	108	55
Degree of roundedness (μm)	105 ± 2.5	32 ± 1.5	48 ± 5	20.5 ± 1.5

Note: The degree of roundedness is obtained by averaging the radii values of the best-fitting quarters of circular arcs to the internal corners of several tube cross sections, obtained by cutting a tube with a precision saw. The uncertainty on degree of roundedness is the standard deviation to the mean value.

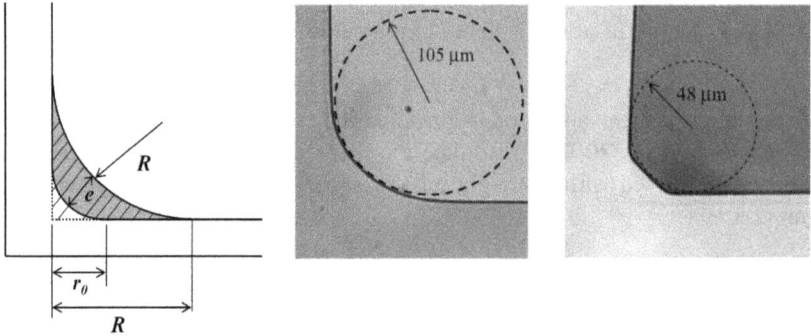

FIGURE 2.4 Left: Radius of curvature R and film thickness e in the tube cross-section plane at a location along the tube where the longitudinal curvature of the liquid–gas interface is negligible compared to the transverse one ($1/R$). The degree of roundness is denoted as r_0. Note that the tube external corners are also rounded but, as it does not play a role here, it is not shown in this sketch. The hatched region is the union of the liquid and solid corners. Middle: Roundedness of the internal corner of tube 1 (see Table 2.1). Right: Roundness of the internal corner of tube 3 (see Table 2.1). The middle and right images are obtained by an optical microscope with a ×20 magnification.

TABLE 2.2
Fluid Properties

	47V5	Heptane	2-Propanol
Density (g cm^{-3})	0.910	0.679	0.781
Surface tension (mJ m^{-2})	19.70	19.66	20.93
Liquid viscosity (kg m^{-1}s^{-1})	4.55	0.387	2.04
Equilibrium vapor mass concentration (kg m^{-3})	/	0.230	0.134
Vapor-air diffusion coefficient (m^{2}s^{-1})	/	7.23×10^{-6}	9.70×10^{-6}

one side of the capillary tube and takes images at a low magnification (spatial resolution ≈ 13 pixels m/m^{-1}). It is used to detect the bulk meniscus location z_2 shown in Figure 2.3. When the liquid films are present up to the tube open end, i.e., $z=z_1=0$, this also amounts to measure the film length $L_f=z_2$.

The second CCD camera provides high magnification images of the tube top (spatial resolution ≈ 654 pixels m/m^{-1}). The optical axis is aligned with one of the tube diagonals so that three corner films out of four are seen on the high magnification images. The focus is made on the two lateral corner films. Imaging processing, based on optical geometry considerations, allows determining the film thickness e. The film thickness measurement is typically performed on one of the two lateral films imaged, at least one tube diameter away from the film tip, i.e., in a region where the film longitudinal curvature can be neglected. Also, the moment when the lateral films depinning from the tube top occurs can be determined by careful inspection of the recorded images. The image acquisition rate is typically 0.05 Hz. For some experiments, it is found that the film tip depinning in each corner does not occur within the same time interval between two given recorded images: after the first film depinning, the following images can still show one or two corner films still present up to the tube top for a short time duration. This situation will be referred to as "differential depinning". However, the difference between the longest corner films imaged and the shortest one is always small, being at most one tube internal diameter.

A typical evolution of the bulk meniscus location z_2 as a function of time is shown in Figure 2.5 for heptane in tube 2. Three distinct periods can be seen. First, as long as $z_2 < 17$ mm, the evaporation rate is roughly constant. Then, around $z_2 \approx 17$ mm, the evaporation rate starts to decrease and this corresponds to the moment when the films depin from the tube's open end, as can be checked visually on the recorded images. For this given experiment, the film tip depinnings do not occur exactly at the same time. The moment when the first and the last film depinning occur is indicated

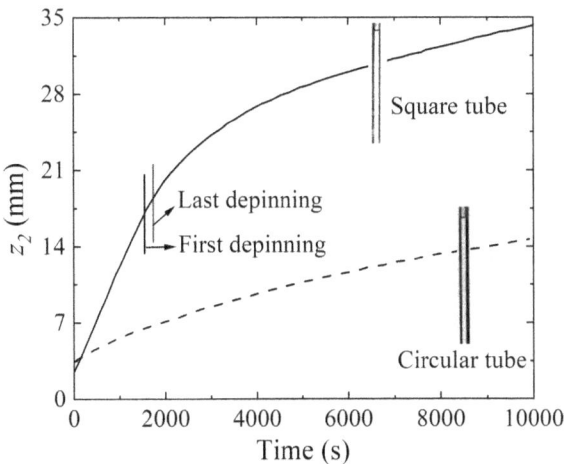

FIGURE 2.5 Bulk meniscus evolution as a function of time for heptane in tube 2. The moment when the first and the last film depinning are observed is indicated on the graph.

in Figure 2.5. Finally, after the film depinning, the film tips recede inside the tube. Consequently, the distance between the film tips and the tube open end increases, increasing the diffusive resistance to mass transfer and thus a continuous decrease of the evaporation rate, as shown in Figure 2.5. Also shown in Figure 2.5 is the variation of z_2 with time for the circular tube. Apparently, the evaporation rate for the square tube is higher than that for the circular tube.

2.3.2　MODEL

In this subsection, a model of evaporation in a tube of square cross section is presented. The predicted results are compared against the experimental results. Here, we consider the slow evaporation in the stagnant air. The phase change is controlled and limited by the vapor diffusion in the surrounding air. The system is assumed to be isothermal, i.e., the temperature variations due to the evaporative cooling effect are neglected. This assumption is supported by the experimental data. Using an infrared thermography technique, the cooling induced by evaporation can be measured. The cooling is located at the top of the tube, where the phase change takes place until the film depinning occurs. The cooling is at most 2.5 K when using 2-propanol in tube 1 and less than 1.5 K for the other cases. This cooling is sufficiently small to neglect the temperature variations.

Based on the analysis of Ransohoff and Radke (1988) for the corner flows, the liquid flow rate along the corner liquid films can be expressed as

$$q_{liq}(z) = \rho_l \frac{A_c R^2}{\beta \mu_l} \left(\frac{dP_l}{dz} - \rho_l g \right) \tag{2.14}$$

where ρ_l is the liquid viscosity, P_l the liquid pressure, and A_c the sum of the areas of the four liquid corner films and of the four "solid" corners, in a cross section at position z:

$$A_c = \lambda R^2 \tag{2.15}$$

with $\lambda = \pi - 4$ for a perfectly wetting liquid. In Eq. (2.14), β is a dimensionless flow resistance depending on the tube corner shape, the contact angle, and the boundary condition at the liquid–air interface. Here, the shear stress on the liquid–gas interface of the corner liquid films can be neglected because the liquid viscosity is much larger than that of gas. The liquid is considered as perfectly wetting on the internal tube wall. The liquid flow rate q_{liq} is positive for the flow from the main gas–liquid interface, i.e., the bulk meniscus illustrated in Figure 2.3, to the open side of the tube.

Various methods have been employed to determine the dimensionless flow resistance as shown in Eq. (2.14). Ransohoff and Radke (1988) solved the corner flow equation numerically using a finite element method and obtained a discrete set of values for the dimensionless flow resistance, depending on the contact angle, the shear stress on the liquid–vapor interface, and the degree of roundedness. An analytical formula is proposed in Zhou et al. (1997) using two classical approaches: the hydraulic diameter approach and the thin film flow theory. In the two aforementioned studies, the dimensionless flow resistance, which is denoted by β, is unbounded and diverges when r_0/R is close to 1. A third approach, initiated by Ayyaswamy et al. (1974) and later revisited by Chen et al. (2006) notably in the case of rounded corner, has shown that the aforementioned divergence of the hydraulic resistance disappears when using a proper scaling of the problem. Using the notations and the results of Chen et al. (2006), the dimensionless flow resistance β can be expressed as

$$\beta = \frac{f^2}{F_i}\frac{1+T_c^2}{\left(1-r_0^*\right)^2 T_c^2}\left\{1-\left[\frac{r_0^*}{h^*\left(1-r_0^*\right)+r_0^*}\right]^2\right\}^{-1}$$

(2.16)

where $f = \sin\alpha \, / \left(\cos\alpha - \sin\alpha\right)$, $T_c = \tan\alpha + f\delta r_0^* \, / \left(1-r_0^*\right)$, $\delta = \pi \, / \, 2 - \alpha - \theta$, $h^* = \left(R - r_0\right)/\left(R_{bm} - r_0\right)$, α is the half-angle of the corner. The variable $F_i = W \times F\left(\alpha, \theta, h^*, r_0^*\right)$, which can be found in Chen et al. (2006).

The liquid pressure of corner film at z is related to the gas pressure P_g by Laplace's law:

$$P_l(z) = P_g - \frac{\sigma}{R(z)}$$

(2.17)

where σ is the surface tension and $R(z)$ is the radius of the curvature of the gas–liquid interface in the plane perpendicular to the tube axis. Here, the curvature along the tube axial direction is neglected. As the hydrostatic pressure drop over the vertical extension of the bulk meniscus is negligible compared to the capillary pressure jump at the bulk meniscus, the total curvature of the bulk meniscus can be taken as the purely capillary solution for the meniscus curvature:

$$R_{bm} = \frac{d}{2\chi}$$

(2.18)

where χ is a dimensionless curvature, which is a function of both the contact angle and the critical contact angle above which no corner film exists. For a perfectly wetting liquid in a square capillary tube, $2\chi \approx 3.77$ (Wong et al., 1992). In the present modeling, it is assumed that the purely transverse radius of curvature of the corner films tends toward R_{bm} when the corner films match to the bulk meniscus, i.e., when z approaches z_2. Here, the film length L is obtained experimentally by measuring the bulk meniscus bottom position z_2. Consequently, the predictions of film lengths from the model come down to neglecting the vertical extension of the bulk meniscus compared to the film length. This is reasonable for the elongated films introduced here.

Combining Eqs. (2.14) and (2.17) leads to

$$q_{liq}(z) = \rho_l \frac{\lambda R^2}{\beta \mu_l}\left(\gamma R^{-2}\frac{dR}{dz} - \rho_l g\right)$$

(2.19)

Due to evaporation along the films, the liquid flow rate decreases up to the film tips. As in Yiotis et al. (2004), the resulting vapor diffusion into the tube is taken into account in the modeling. Here, the mass transfer between the liquid film and the gas phase inside the tube is simply modeled by introducing a mean vapor mass concentration $\bar{c}$, which is spatial average of the vapor mass concentration over the tube cross section region occupied by the gas $A_g = d^2 - \lambda R^2$, and assuming the gas mixture as being dilute such that the evaporation mass rate per unit of length inside the tube Q_{ev} is given by

$$Q_{ev} = 2\pi DhR\frac{c_e - \bar{c}}{d/2} = 2\pi DhR\frac{c_e - \bar{c}}{\chi R_{bm}} \tag{2.20}$$

where c_e is the equilibrium vapor mass concentration at the liquid–gas interface, D is the vapor molecular diffusion coefficient, and $h = 4$ is a dimensionless coefficient of mass transfer provided by three-dimensional computations of the steady diffusion of vapor in the gas phase, inside a square capillary tube partially filled with a perfectly wetting and volatile liquid (Camassel et al., 2005).

The diffusional mass flux in the gas phase q_{vap} is given by

$$q_{vap}(z) = DA_g\frac{d\bar{c}}{dz} \tag{2.21}$$

Here, q_{vap} is positive if vapor diffuses from the main gas–liquid interface to the open side of the tube. The evaporation rates expected are such that the characteristic drying time $\tau_{dry} = d\left(dz/dt^{-1}\right)$ is much larger than the characteristic diffusion time $\tau_{diff} = d^2/D$, so that vapor diffusion in the tube can be considered as being quasi-steady. Also, the characteristic time for the interface shape to reach a steady state is much smaller than τ_{dry}, so that the interface shape variation can be considered as being quasi-steady (Yiotis et al., 2004). Consequently, one obtains from mass conservation

$$Q_{ev} = \frac{dq_{liq}}{dz} = -\frac{dq_{vap}}{dz} \tag{2.22}$$

The total mass flux is denoted as q_{tot}, $q_{tot} = q_{liq} + q_{vap}$, and is constant along z.

If we focus on the moment when the film tips depinning occurs from the tube top, the following two boundary conditions have to be used:

$$q_{vap}(0) = q_{tot} \tag{2.23}$$

$$q_{liq}(0) = 0 \tag{2.24}$$

Here, the gravity force is neglected over the bulk meniscus vertical extension. In addition, it is assumed that the liquid flow rate is small enough to consider that the capillary pressure jump at the bulk meniscus is constant and equal to its static value. Also, the vapor mass concentration on the liquid–gas interface is assumed to be constant and equal to its equilibrium value. The two resulting boundary conditions read

$$R(z_2) = R_{bm} \tag{2.25}$$

$$\bar{c}(z_2) = c_e \tag{2.26}$$

The above equations are made dimensionless using the following variables:

$$R^* = \frac{R}{R_{bm}}, \quad z^* = \frac{z}{L}, \quad c^* = \frac{\bar{c}}{c_e}$$

(2.27)

Combining Eqs. (2.19), (2.20), and (2.22), a second-order nonlinear ordinary differential equation on R^* is obtained:

$$R^* \frac{d^2 R^*}{dz^{*2}} + \left(2 - \frac{R^*}{\beta} \frac{d\beta}{dR^*}\right)\left(\frac{dR^*}{dz^*}\right)^2 - \mathrm{Bo}\,\epsilon^{-1} R^{*2}\left(4 - \frac{R^*}{\beta} \frac{d\beta}{dR^*}\right)\frac{dR^*}{dz^*}$$

$$= \epsilon^{-2} \beta \mathrm{Cah}\left(1 - c^*\right)$$

(2.28)

where $\epsilon = R_{bm}/L$, and Ca is the capillary number:

$$\mathrm{Ca} = \frac{2\pi\mu_l D c_e}{R_{bm}\rho_l \lambda\gamma}$$

(2.29)

The Bond number Bo is

$$\mathrm{Bo} = \frac{\rho_l g R_{bm}^2}{\gamma}$$

(2.30)

Combining Eqs. (2.20)–(2.22), another second-order ordinary differential equation on c^* is obtained:

$$\left(4\chi^2 - \lambda R^{*2}\right)\frac{d^2 c^*}{dz^{*2}} - 2\lambda R^* \frac{dR^*}{dz^*}\frac{dc^*}{dz^*} = -2\pi\epsilon^{-2} hR^*\left(1 - c^*\right)$$

(2.31)

The boundary conditions, Eqs. (2.23)–(2.26), become

$$q_{vap}^*(0) = q_{tot}^*$$

(2.32)

$$q_{liq}^*(0) = 0$$

(2.33)

$$R^*(1) = 1$$

(2.34)

$$c^*(1) = 1$$

(2.35)

where the flux is made dimensionless by a reference evaporation rate $q_{ref} = R_{bm} D c_e$ and with $q_{tot}^* = q_{liq}^* + q_{vap}^*$. To close this set of equations, the following boundary condition is added:

$$R^*(0) = r_0^*$$

(2.36)

Equation (2.36) indicates that the film tips are at depinning.

The two coupled Eqs. (2.28) and (2.31), along with the corresponding boundary conditions, form a boundary value problem with one unknown parameter ϵ. An iterative scheme is used to solve this problem. Equations (2.28) and (2.31) are solved successively. At each iteration, Eq. (2.28) on R^*, with boundary conditions (Eqs. 2.33, 2.34, and 2.36) and Eq. (2.31) on c^*, with boundary conditions (Eqs. 2.32 and 2.35) are solved. The parameter ϵ is calculated in the same time as R^*. As β and $\dfrac{R^*}{\beta}\dfrac{d\beta}{dR^*}$ diverge when $R^* \rightarrow r_0$, the boundary condition Eq. (2.36) is in fact applied as $R^*(0) = x r_0^*$ with $x = 1.001$ which, after several tests, appeared to be sufficient to obtain a converged solution. Note that in the present numerical procedure, the value of q_{tot} is used as an input. Three-dimensional numerical simulations of the vapor diffusion problem in the capillary tube, from the meniscus surface to the stagnant ambient air, could be used to compute the evaporation rate as a function of the bulk meniscus position and of the full meniscus surface shape. However, this numerical procedure is heavy, and for simplicity, the values of q_{tot} will be obtained experimentally.

In Figure 2.6, the dimensionless film thickness $e^* = \left(R^* - r_0^*\right)/\left(1 - r_0^*\right)$ is shown as a function of z^* for several values of Bond and capillary numbers. The value of q_{tot}^* used is typical of the values measured at the depinning in the experiments in Section 3. The corresponding dimensionless film length at depinning $\epsilon^{-1} = L/R_{bm}$ is given in the legend. In Figure 2.6a, the film thickness profiles are shown for several Bond numbers at Ca $= 5 \times 10^{-6}$. The effect of gravity on such profiles is clearly visible: the corner film profile is all the more thin and short that the Bond number is large. In Figure 2.6b, the film thickness profiles are for several capillary numbers at Bo $= 5 \times 10^{-3}$ for the same set of parameters as in Figure 2.6a. The viscous effects clearly limit the film length.

The effect of r_0 on the film length at depinning is displayed in Figure 2.7 for different values of the capillary number at a fixed value of $|q_{tot}|$ and of the Bond number. The corner films maximal extension (i.e., obtained at the moment when the depinning from the top of the tube occurs) is found to decrease with r_0. This is an important effect. For instance, for Ca $= 10^{-6}$, the film length at depinning is divided

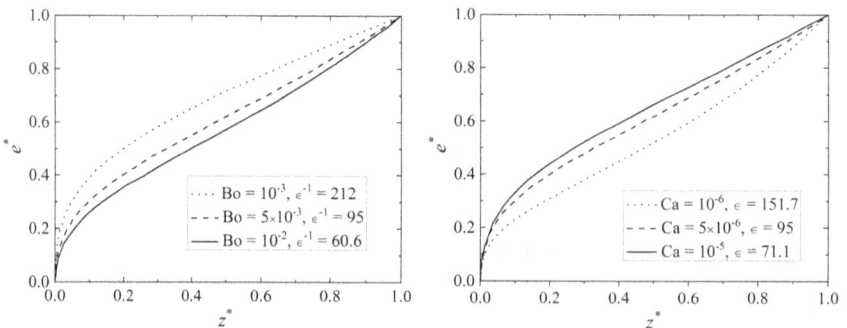

FIGURE 2.6 Dimensionless film thickness $e^* = (R^* - r_0^*)/(1 - r_0^*)$ as a function of the dimensionless z^* coordinate. $r_0^* = 0.4$, $|q_{tot}^*| = 4.5$, and the results of Zhou et al. (1997) are used for β. (a) Ca $= 5 \times 10^{-6}$ and (b) Bo $= 5 \times 10^{-3}$.

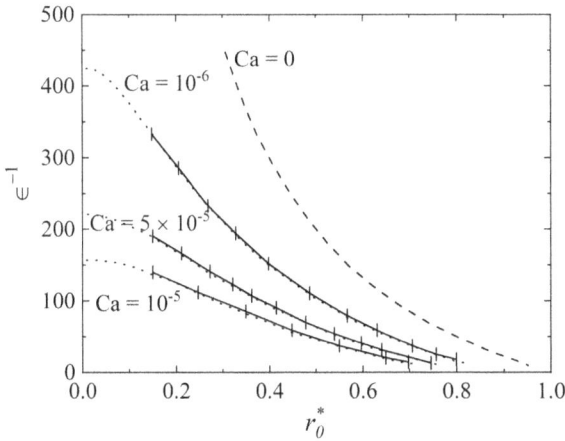

FIGURE 2.7 Dimensionless liquid film length $\varepsilon^{-1} = L/R_{bm}$ as a function of the dimensionless degree of roundedness r_0^* for various Ca numbers. Solid and dashed lines correspond to the results of complete model (Eqs. (1.28–1.36)) and the dotted lines to those of the simplified model (Eqs. (1.36–1.38)). For all the curves, Bo $= 5 \times 10^{-3}$, $|q_{tot}^*| = 4.5$, and the results of Zhou et al. (1997) were used for β.

by more than a factor of 2 between the case of a perfectly sharp corner and the case of a rounded tube with $r_0^* = 0.4$. For any value of r_0^*, it is important to note that the film extension is finite and much smaller than the one that is obtained in a purely hydrostatic case, when Ca$=0$, which highlights how the viscous effects induced by the liquid flow toward the tube top limit the film length. When $r_0 \rightarrow 1$, the film length at depinning decreases: as the present model is no longer valid when the film becomes so short that the effect of the longitudinal curvature on the film length cannot be neglected, no results for $\varepsilon^{-1} < 10$ are shown in Figure 2.7. The film length depends weakly on the value of the dimensionless mass transfer coefficient h. For instance, for Ca $= 10^{-5}$, Bo $= 5 \times 10^{-3}$, $r_0^* = 0.4$, and $|q_{tot}^*| = 4.5$, ε^{-1} varies between 71.2 and 70.6 when h is varied between 2 and 6. This weak dependence of the film length on h is due to the fact that the extension within the tube over which the vapor flow rate q_{vap} is nonzero is always much shorter than the film length, for all the tested values of h (see Figure 2.8). The sensitivity of the obtained film length to the fixed value of q_{tot} chosen for the computations at a given Ca and r_0^* is also shown in Figure 2.7. The results obtained for $|q_{tot}| = 4.5 \pm 0.45$ are plotted as errors bars in Figure 2.7. This clearly displays that the film length is mainly controlled by the tube roundedness. For instance, at Ca $= 5 \times 10^{-6}$ and $r_0^* = 0.3$, a 10% variation in $|q_{tot}|$ results in a variation of $\approx 3.5\%$ for ε^{-1}, whereas a 10% variation in r_0^* results in a variation of $\approx 10\%$ for ε^{-1}.

As can be seen in Figure 2.8a, where the vapor mass concentration profiles in the tube are shown for the same set of parameters as in Figure 2.6a, the gas phase is saturated over much of the film length: the vapor concentration gradients are only significant near the tube entrance region for $0 < z^* < 0.1$. Consequently, the phase change occurs preferentially at the film tip: the vapor mass flux is nonzero only in the near tube entrance region, as seen in Figure 2.8b, which shows the absolute value

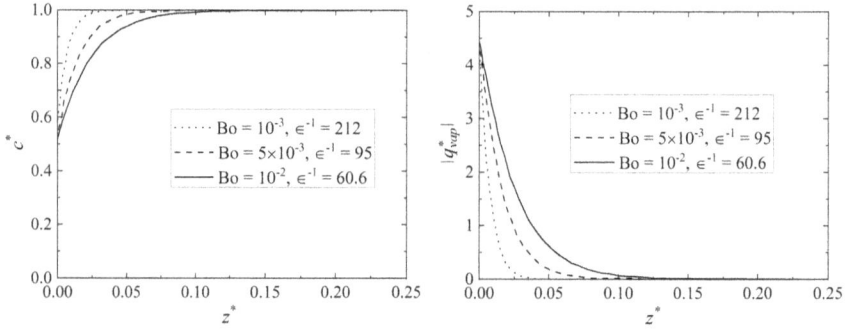

FIGURE 2.8 (a) Dimensionless vapor mass concentration c^* as a function of z^*. (b) Absolute value of the dimensionless vapor mass flux r^*_{vap} as a function of z^*. These results are obtained using $Ca = 5 \times 10^{-6}$, $r^*_0 = 0.4$, $|q^*_{tot}| = 4.5$, and the results of Zhou et al. (1997) for β.

of the dimensionless vapor mass flux q^*_{vap} as a function of z^*. In the following, a simpler model is proposed in which vapor diffusion is neglected in the tube: the air is considered to be saturated in vapor inside the tube, so that $q^*_{vap} = 0$ at any $z^* > 0$. Consequently, the liquid flow rate is assumed to be constant along the tube. This model is similar to the one proposed in Chauvet et al. (2010). It can predict, as shown below, the evolution of the film thickness during the whole evaporation process and not only at the moment when the film depinning occurs.

With the additional assumption on the saturation of air with vapor inside the tube, the problem amounts to a single differential equation, obtained by rewriting Eq. (2.19) as

$$\frac{d\tilde{z}}{dR^*} = \left(\frac{Ca}{2\pi} q^*_{tot} \beta R^{*-2} + Bo R^{*2} \right) \tag{2.36}$$

where now $\tilde{z} = z/R_{bm}$. Using q_{tot} as an input and solving Eq. (2.36) using a finite difference scheme with the boundary condition

$$\tilde{z} = 0 \text{ at } R^* = x r^*_0 \text{ with } x = 1.001 \tag{2.37}$$

leads to a prediction of the film length at depinning, which is then $\epsilon^{-1} = \tilde{z}(R^* = 1)$. Equation (2.36) can also be solved using the following boundary condition:

$$\tilde{z} = \tilde{z}_2 \text{ at } R^* = 1 \tag{2.38}$$

using the measured q_{tot} and film length as inputs in order to get the film thickness profile $R^*(\tilde{z})$.

The predictions of this simple model as far as the film length at depinning is concerned, i.e., using boundary condition in Eq. (2.38), are shown as dotted lines in Figure 2.7. The simple model provides film lengths very close to the complete

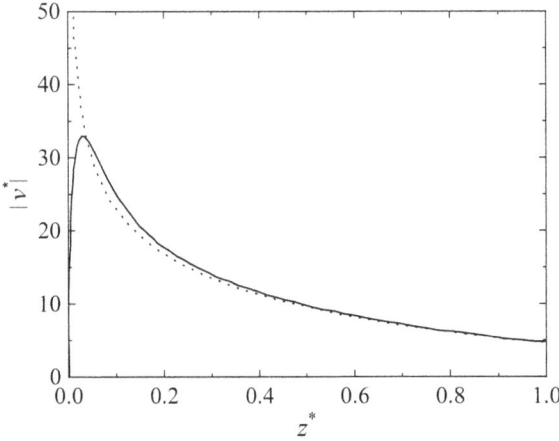

FIGURE 2.9 Absolute value of the dimensionless liquid mean velocity $|v^*|$ as a function of z^*. These results are obtained using Bo = 5×10^{-2}, Ca = 5×10^{-6}, $r_0^* = 0.4$, $|q_{tot}^*| = 4.5$, and the results of Zhou et al. (1997) for β. Solid line: complete model; dotted line: simplified model.

model predictions (solid lines in Figure 2.7) for a wide range of capillary numbers, Bond numbers, and total evaporation rate values. It can be noted that the film lengths given by this simplified model are always slightly lower than those obtained with the previous approach (the maximal difference for the results presented in Figure 2.7 is 2.5 R_{bm} for the Ca = 10^{-6} case, at $r_0^* = 0.15$). Indeed, in the simple model, the liquid flow rate is constant along the entire film length. Therefore, it can be argued that in this case, the liquid pressure gradient dp_l/dz, in the film tip region, is larger than the one obtained using the previous model. As $dp_l/dz \propto -d[1/R(z)]/dz$, the radius of curvature gradient is then lower in the previous model than in the simplified model, leading to larger film length at depinning. It is also important to notice that even if this simplified model is in good agreement with the previous one as far as film length at depinning is concerned, it predicts a divergence of the dimensionless liquid mean velocity at the film tip v^* (see Figure 2.9), where

$$v^* = \frac{q_{lid} / [\rho_l \lambda (R^2 - r_0^2)]}{q_{ref} / [\rho_l \lambda (r_{bm}^2 - r_0^2)]} = q_{lid}^* \frac{1 - r_0^{*2}}{R^{*2} - r_0^{*2}} \tag{2.39}$$

Indeed, since the liquid flow rate is constant along the film and since the liquid cross section at the film tip is 0, a divergence of v^* is found at the film tip. On the contrary, as the complete model takes into account evaporation along the films, it predicts a more realistic zero liquid mean velocity at the film tip (see Figure 2.9).

It should be noted that evaporation of the straight tube of polygonal cross section other than square, i.e., triangular and hexagonal, also has been studied (Prat, 2007). A detailed study of evaporation in the entrance region of a channel with corner films can be found in Keita et al. (2016).

2.4 EVAPORATION OF INTERCONNECTED STRAIGHT TUBES WITH VARIOUS SIZES

In the foregoing sections, drying in a single tube was discussed. As a matter of fact, the void space in a porous material is composed of lots of pores connected to each other (these pores have different sizes). As a further step for the understanding of evaporation in porous media, we consider in this subsection the evaporation in several interconnected straight tubes with different sizes (Metzger & Tsotsas, 2005). Figure 2.10a shows five parallel capillaries (not to scale) that are at first completely filled with water. The left end of the capillaries is open for evaporation; the right end is closed. The positions $x=0$ and L correspond to the surface and the center (or sealed back) of the porous medium, respectively. The length of the capillaries is $L=0.1$ m; their radii are $r_i=110$, 105, 100, 95, and 90 nm.

At the beginning of the drying process, the meniscus of the biggest capillary retreats from the surface (position s_1), while the others remain stationary due to capillary forces, Figure 2.10b. Liquid water is pumped out of the biggest capillary through the other four capillaries to the surface. If the pressure drop in the full capillaries

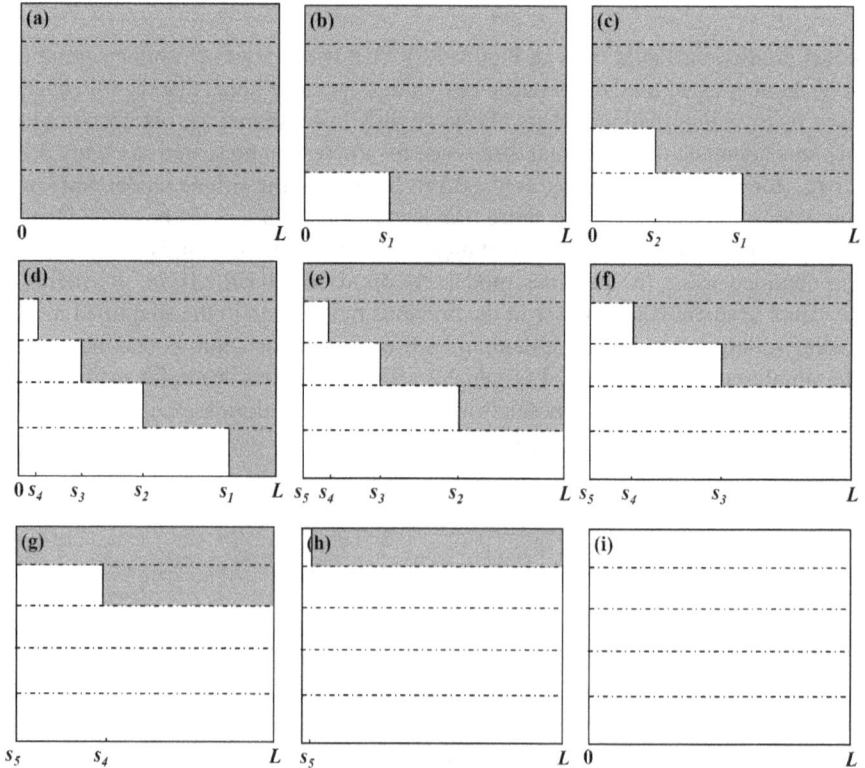

FIGURE 2.10 Filling of capillaries at distinct times during isothermal drying (the evaporation surface is on the left, shown are all meniscus position for which $0<s_i<L$).

Δp_f (due to friction) equals the difference in capillary pressure Δp_c, another meniscus retreats from the surface (Figure 2.10c) until only one meniscus is left at the surface, Figure 2.10d. This marks the end of the first evaporation period. Up to this point, part of the surface is still wet, and the evaporation rate may be set equal to that for a completely wet surface. This is a good approximation as long as the radius of the wet patches is small compared to the thickness of the boundary layer and as long as the fraction of wet surface is not too small (Schlunder, 1988). (Note that for a different capillary geometry, some capillaries may empty before the end of the first drying period.)

Then, the meniscus of the smallest capillary (s_5) moves back from the surface (the distance is too small to be seen in Figure 2.10). This introduces an additional mass transfer resistance for evaporation so that the drying rate is reduced. As a consequence, the frictional forces in the capillaries decrease, whereas the capillary forces stay constant. Hence, the distance between the menisci may increase and one capillary after the other will empty (Figure 2.10e–g) while the last meniscus is still relatively close to the surface (Figure 2.10h). Then, the evaporation rate will decrease even stronger as the evaporating meniscus retreats without supply from other capillaries. We assume that all empty capillaries are saturated with vapor up to the last meniscus (no resistance between capillaries) and that consequently the whole cross section serves as an evaporation area. Evaporation is completed at a small but non-zero evaporation rate (Figure 2.10i). Residual moisture in the form of adsorbed water is not accounted for.

Evaporation of water is described as a diffusive vapor flux given by

$$q_{vap} = -\left(\frac{1}{D_s} + \frac{s_n}{D_{AB}}\right)^{-1} \frac{pM_v}{R_uT} \ln \frac{(p - p_{v,\infty})}{(p - p_v^*)} \tag{2.40}$$

where D_s is the mass transfer coefficient at the surface, s_n is the position of the meniscus in the smallest capillary, D_{AB} is the diffusion coefficient of vapor in air, A represents water vapor, B represent air, p is the total pressure, M_v is the molar mass of vapor, R_u is the universal gas constant, T is the temperature (in K), $p_{v,\infty}$ is the vapor pressure in the bulk drying air, and p_v^* is the saturation vapor pressure. The Knudsen effect in vapor diffusion is neglected, which is reasonable as long as the radius of the capillaries is bigger than the mean free path of the vapor molecules (≈ 40 nm). The reduction of vapor pressure above the curved liquid surfaces (Kelvin effect) is not taken into account, which is a good approximation for $r > 10$ nm. We take $D_s = 0.1$ m/s as well as atmospheric pressure throughout this chapter, and $T = 20°C$, $p_{v,\infty} = 0$, and $L = 0.1$ m for the isothermal case. The water that evaporates must also flow as liquid through the capillaries. The number of liquid-filled capillaries varies according to the position in the product. The positions of the menisci s_i give a natural subdivision of space. Between two consecutive meniscus positions, the pressures that must balance are the difference in capillary pressure (due to different radii):

$$\Delta p_{c,i} = 2\gamma\left(\frac{1}{r_{i+1}} - \frac{1}{r_i}\right) \text{ for } i = 1, ...,n - 1 \tag{2.41}$$

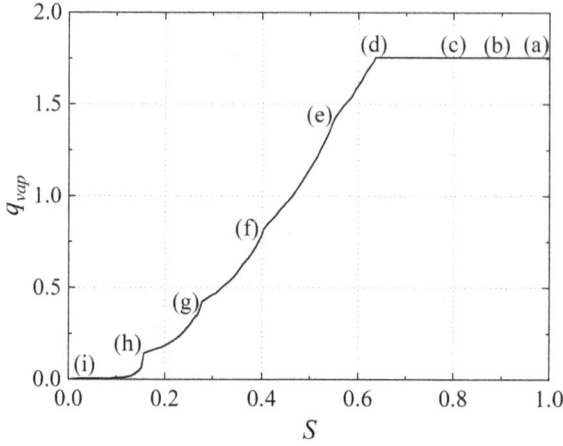

FIGURE 2.11   Drying curve for capillaries in Figure 1.10 (the distinct times are shown on the curve).

The pressure drop due to friction (Poiseuille flow):

$$\Delta p_{f,i} = 8\mu \frac{\dot{m}_v}{\rho_w} \left(\sum_{j=1}^{n} N_j \pi r_j^2 \right) \left(\sum_{j=i+1}^{n} N_j \pi r_j^2 \right)^{-1} (s_i - s_{i+1}) \qquad (2.42)$$

where γ is the surface tension, μ is the dynamic viscosity, ρ_w is the density of water, and N_j denotes the number of capillaries for each class. (The first sum is over all capillaries and gives the total surface for evaporation; the second sum is only over liquid-filled capillaries to account for friction.) Note that we have assumed a zero contact angle, i.e., complete wetting of walls, in Eq. (2.46).

The pressure balance between Eqs. (2.41) and (2.42) defines the distances of the menisci $s_i - s_{i+1}$ as a function of the drying rate, while Eq. (2.40) gives the absolute position of the last meniscus s_n. Hence, the model is complete, and integration over time can yield moisture profiles and the drying curve.

Figure 2.11 shows the evaporation curve for the complete process. We use saturation S to describe the liquid saturation because volume fraction and density of the solid phase are not relevant in this model. As can be seen, the evaporation rate remains almost constant when the saturation $S > 0.62$. For $S < 0.62$, the evaporation rate reduces with the decrease of S.

2.5   SUMMARY

The mass transfer driven evaporation in capillary tubes is introduced. We start from evaporation in a single capillary tube with circular cross section. For such case, the corner liquid films are absent. A theoretical model is developed for the vapor diffusion inside the tube. The predicted variation of the location of the meniscus inside tube during evaporation is in good agreement with the experimental results. Then, evaporation in the straight capillary tube with square cross section is studied. In this case, the liquid films trapped by capillarity in the corners of the square cross-section tube, i.e., the so-called corner liquid films, have a great impact on the evaporation kinetics. The corner liquid films provide paths for the liquid between the receding bulk meniscus and the entrance of the tube. The capillary pumping of the volatile liquid from the bulk meniscus to the tube open end where it evaporates is the key point in explaining the much faster evaporation rate obtained in a square capillary tube (compared to a circular cross-section tube). A model is developed to depict the evaporation in the capillary tube with the corner liquid films. The tube corner roundedness has a great influence on the evaporation kinetics. Finally, evaporation from multiple capillary cylindrical tubes is introduced. Due to the effects of capillary forces, during evaporation, the liquid in bigger capillaries can be transported to smaller capillaries, which keeps the menisci in smaller capillaries near the outlet of the capillaries and thus maintains the first evaporation period. The kinetics of evaporation in straight capillary tubes presented in this chapter is key to understand in detail the evaporation in porous media composed of interconnected pores (tubes) of various sizes.

REFERENCES

Ayyaswamy, P.S., Catton, I., Edwards, D.K. 1974. Capillary flow in triangular grooves. *Journal of Applied Mechanics* 41: 332–336.

Bird, R.B. Stewart, W.E. Lightfoot, E.N. 1960. *Transport Phenomena*. John Wiley & Sons, New York, 413.

Camassel, B. Sghaier, N. Prat, M. Nasrallah, S.Ben. 2005. Evaporation in a capillary tube of square cross section: application to ion transport. *Chemical Engineering Science* 60: 815–826.

Chauvet, F. Cazin, S. Duru, P. Prat, M. 2010. Use of infrared thermography for the study of evaporation in a square capillary tube. *International Journal of Heat and Mass Transfer* 53: 1808–1818.

Chauvet, F. Duru, P. Geoffroy, S. Prat, M. 2009. Three periods of drying of a single square capillary tube. *Physical Review Letters* 103: 124502.

Chen, Y. Weislogel, M.M. Nardin, C.L. 2006. Capillary-driven flows along rounded interior corners. *Journal of Fluid Mechanics* 566: 235–271.

Concus, P. Finn, R. 1969. On the behavior of a capillary surface in a wedge. *Proceedings of the National Academy of Sciences of the United States of America* 63: 292.

Keita, E., Koehler, S.A., Faure, P., Weitz, D.A., Coussot, P. 2016, Drying kinetics driven by the shape of the air/water interface in a capillary channel. *The European Physical Journal E* 39: 23.

Metzger, T. Tsotsas, E. 2005. Influence of pore size distribution on drying kinetics: a simple capillary model. *Drying Technology* 23: 1797–1809.

Prat, M. 2007. On the influence of pore shape, contact angle and film flows on drying of capillary porous media. *International Journal of Heat and Mass Transfer* 50: 1455–1468.

Ransohoff, T.C. Radke, C.J. 1988. Laminar flow of a wetting liquid along the corners of a predominantly gas-occupied noncircular pore. *Journal of Colloid and Interface Science* 121: 392–401.

Schlunder, E.-U. 1988. On the mechanism of the constant drying rate period and its relevance to diffusion controlled catalytic gas phase reactions. *Chemical Engineering Science* 43: 2685–2688.

Whitaker, S. 1991. Role of the species momentum equation in the analysis of the Stefan diffu-
    sion tube. *Industrial & Engineering Chemistry Research* 30: 978–983.
Wong, H. Morris, S. Radke, C. 1992. Three-dimensional menisci in polygonal capillaries.
    Journal of Colloid and Interface Science 148: 317–336.
Yiotis, A.G. Boudouvis, A.G. Stubos, A.K. Tsimpanogiannis, I.N. Yortsos, Y.C. 2004. Effect
    of liquid films on the drying of porous media. *AIChE Journal* 50: 2721–2737.
Zhou, D. Blunt, M. Orr Jr., F.M. 1997. Hydrocarbon drainage along corners of noncircular
    capillaries. *Journal of Colloid and Interface Science* 187: 11–21.

3 Pore Scale Experiments on Evaporation from Porous Media

Rui Wu
Shanghai Jiao Tong University

CONTENTS

3.1   INTRODUCTION

During evaporation of a porous medium, the initially liquid-filled pores are gradually occupied by gas. This gas–liquid interface displacement is influenced by the interplay of the viscous forces, gravity, thermal gradient, pore structure, and wettability. The gas–liquid interface displacement determines the phase distribution in porous media and hence the evaporation kinetics. Accurate description of the gas–liquid interface displacement is therefore of vital importance. Although much progress has been made on evaporation in porous media, unveiling the dynamics of gas–liquid interface displacement in porous materials is still a challenge. This is due to the fact that real porous materials are often opaque and have complex pore structures.

Visualization experiments with microfluidic pore networks composed of regular pores, which enable precise visualization of the gas–liquid interface, are proven to be a useful approach for understanding two-phase transport phenomena in porous

DOI: 10.1201/9781003011811-3

media from the pore-scale perspective (Wu et al., 2016; Odier et al., 2017; Edery et al., 2018). In this chapter, evaporation in the microfluidic pore networks is introduced. As we discussed in Chapter 2, the corner liquid films play an important role in evaporation of capillary tubes. Thus, evaporation of the microfluidic pore networks with and without the corner liquid films is presented in this chapter so as to disclose the effects of the corner liquid films on evaporation in porous media. During evaporation in porous media, many gas–liquid interfaces, i.e., the so-called menisci, are formed. These menisci can interact with each other. Because of the interaction between two menisci, the refilling of liquid into the pores occupied by gas also can be observed. As we discussed later, this refilling is attributed to the capillary instability induced by the heterogeneity in wettability and structure of the pore network. Revealing this capillary instability effect definitely enriches our understanding of evaporation in porous media.

In what follows, evaporation in the PDMS (Polydimethylsiloxane)-based microfluidic pore network is introduced. For evaporation in PDMS pore network, the corner liquid films cannot be formed. To reveal the role of the corner liquid films, evaporation in the silicon-glass based microfluidic pore network is introduced in Section 3.3. The capillary instability due to the interaction of menisci is introduced in Section 3.4. Finally, the conclusion is drawn in Section 3.5.

3.2 EVAPORATION IN THE PDMS-BASED MICROFLUIDIC PORE NETWORK WITHOUT CORNER LIQUID FILMS

In this section, evaporation in a PDMS-based microfluidic pore network is introduced (Wu et al., 2016). PDMS is transparent and water repellent. For evaporation in a PDMS-based pore network, the apparent contact angle between the air–water interface and the water–PDMS surface is about 69°. This means that the corner liquid films cannot exit in the pore with rectangular cross-section in the pore network. The condition for the formation of a liquid film in a pore corner, as stated in Mahmud and Naguyen (2006), is $\alpha + \theta < 90°$, where α is half of the corner angle (e.g., for cuboid ducts, $\alpha = 45°$) and θ is the apparent contact angle between the gas–liquid interface and the liquid–solid interface ($\theta = 69°$ for water evaporation in PDMS pore network).

3.2.1 EXPERIMENTAL SETUP AND IMAGE PROCESSING

It should be noted that PDMS is permeable to water. To reduce water loss due to permeation, both the top and bottom sides of the pore network (perpendicular to the pore depth direction) are covered by a glass sheet of 1 mm thick, Figure 3.1a. We find that if the pore network is not covered by the glass sheet, the total evaporation time will be about three times shorter. To eliminate the viscosity effect on the two-phase flow in the network during evaporation, only one of the pores in the microfluidic pore network is open to the environment through a long tube, Figure 3.1a. The purpose of this long tube is twofold. First, it controls the drying rate at a low level, which guarantees that the two-phase flow in the network is dominated by capillary forces and the viscosity effect can be neglected. Second, this tube gives an explicit boundary

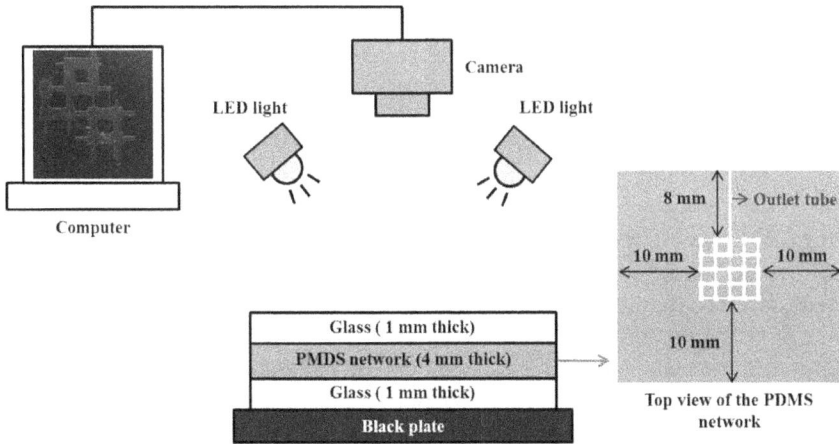

FIGURE 3.1   (a) Schematic of the experimental setup used to record evaporation of a PDMS-based pore network and (b) structure of the pore network.

condition for the mass transfer between the environment and the pore network. To relieve the gravity effect, the pore network is horizontally placed during the evaporation experiment.

The pore network consists of regular pore bodies and pore throats. In the plane shown in Figure 3.1b, the pore bodies are square and the pore throats are rectangular. All the pore bodies have the same side length of $l=1$ mm, and the distance between the centers of two neighboring pore bodies is $a=2$ mm, which means that the pore throat length is $l_t = 1$ mm. The pore throat widths (w) are randomly distributed according to a discrete uniform probability density function in the range of [0.14–0.94] mm with an increment of 0.02 mm so as to eliminate effects of the fabrication uncertainty (0.01 mm). To guarantee that each pore throat has a different width, the size of the pore network is limited to 5×5 pore bodies, as shown in Figure 3.1b, where the numbers are the throat widths (the unit is mm). All the sides of the pore network are closed except that the middle pore at one side of the pore network is open to the environment through a tube of 0.5 mm wide and $\delta = 8$ mm long (for convenience, this tube is called the outlet tube). All pores in the pore network and the outlet tube have the same depth of $h = 0.1$ mm in the direction perpendicular to the plane shown in Figure 3.1b.

Initially, the pore network is immersed into de-ionized water contained in a plastic cylinder depressurized by a vacuum pump so as to saturate the pore network with water. Then, the pore network is horizontally placed on a black plate in a chamber with almost constant temperature (23.8°C ± 1°C) and relative humidity (24% ± 2%). Time variation of the liquid distribution in the pore network is recorded by a camera (Nikon D810, Japan) equipped with a macro lens (AF-S VR Micro-Nikkor 105 mm f/2.8G IF-ED, Japan). The camera is controlled by a computer through the software of Camera Control Pro 2, and the images are acquired at a time interval of 5 minutes. The captured images have a resolution of 7,360×4,912 pixels, which provides a spatial resolution of about 5 μm per pixel.

In our experiment, dye agents are not used in liquid in order to avoid any possible wettability change induced by the presence of dye agents and to prevent the pore blockage due to deposition of the dye agents. For this reason, solid and liquid zones in optical images exhibit a similar color, e.g., Figure 3.2a. To delineate the liquid distribution in the pore network, the optical raw images are processed by using MATLAB. The image processing procedure, as illustrated in Figure 3.2, has the following three steps:

(1) Raw images are true color (e.g., Figure 3.2a and c); for a convenient analysis, these images are converted to gray-scale ones (e.g., Figure 3.2b and d) for which each pixel has a gray value (GV) ranging from 0 to 255; a pixel is black if its GV is 0 and white if its GV is 255. In the following paragraphs, experimental images are referred to as gray-scale images.
(2) Since the pore network remains stationary during the drying experiment, information of positions of solid pixels in all experimental images is the same. Positions of solid pixels can be obtained from the experimental image of the empty network at the end of drying for which there are only two types of pixels: solid and gas pixels, as shown in Figure 3.2d. In this image, solid pixels are darker than gas ones, and pixels at solid–gas interfaces are rather bright. Based on these two features and the pore network structure shown in Figure 3.1b, a pixel in Figure 3.2d is identified as a solid pixel if its GV is lower than a threshold value, which, however, varies spatially since illumination is not uniform. In this way, solid and gas pixels in Figure 3.2d are identified, as illustrated in Figure 3.2e, where black (GV = 0) and white (GV = 225) are the solid and gas pixels, respectively. Based on position

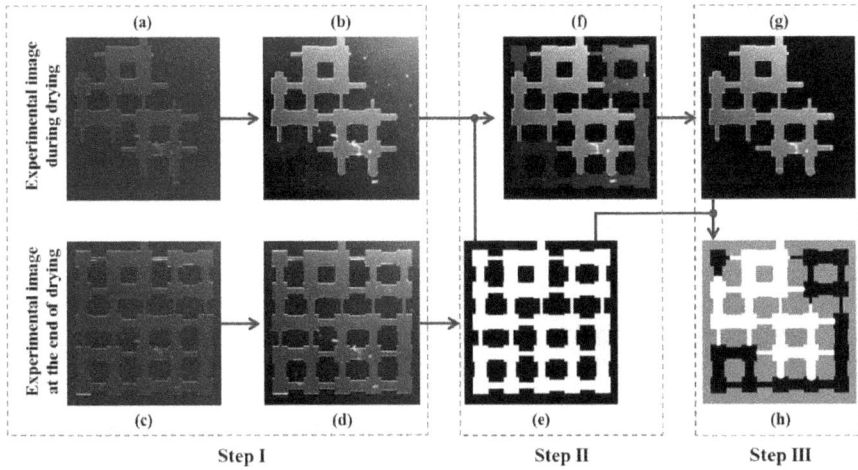

FIGURE 3.2 Procedures to process the experimental images. Experimental image obtained (a) during evaporation (true color image), (b) during evaporation (gray-scale image), (c) at the end of evaporation (true color image), and (d) at the end of evaporation (gray-scale image). (e) Information of position of solid pixels (black) and gas pixels (white) for the image gained at the end of evaporation. (f) Information of positions of solid pixels (black) for the image gained during evaporation. (g) Information of position of solid and liquid pixels (black) for the image gained during evaporation. (h) Liquid distribution for the image gained during evaporation.

information of pixels in Figure 3.2e, solid pixels in experimental images of partially filled networks obtained during drying, e.g., the one shown in Figure 3.2b, can be determined, and their GVs are set to zero. This leads to the image of Figure 3.2f, where black (GV = 0), dark gray, and white gray are the solid, liquid, and gas pixels, respectively.

(3) The liquid distribution in Figure 3.2f is still not clear. To address this issue, we first paint liquid pixels in this figure black using the software of PAINT so as to make GVs of liquid pixels to zero. This results in the image shown in Figure 3.2g for which both solid and liquid pixels have GV = 0, while GVs of gas pixels are higher (> 30). Then, pixels with GV > 30 (i.e., gas pixels) in Figure 3.2g are set to GV = 255, and GV = 125 is applied to the solid pixels in this figure based on pixel information of Figure 3.2e. In this way, we get the image shown in Figure 3.2h, where gray (GV = 125), black (GV = 0), and white (GV = 255) are the solid, liquid, and gas pixels, respectively. Hence, the liquid distribution in the pore network is clearly obtained, and the liquid saturation is defined as the ratio of the number of liquid pixels to the number of liquid and gas pixels.

3.2.2 EXPERIMENTAL RESULTS

The variation of the liquid distribution in the pore network during evaporation is shown in Figure 3.3. The variation of the liquid saturation in the course of evaporation

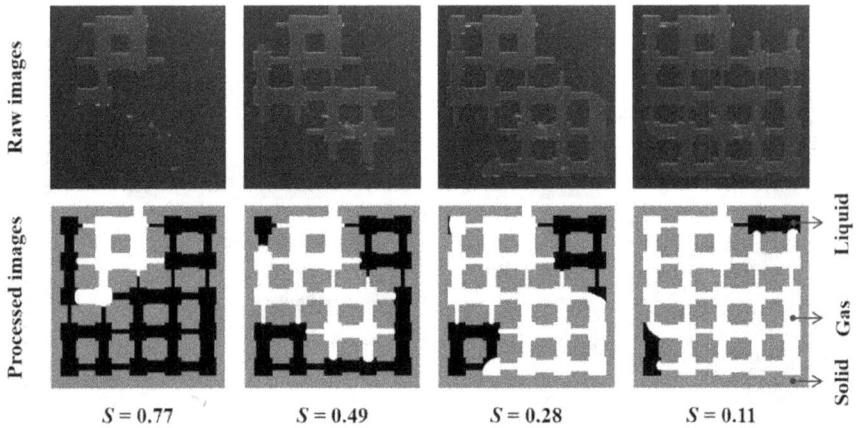

FIGURE 3.3   Variation of the liquid distribution in the pore network during evaporation.

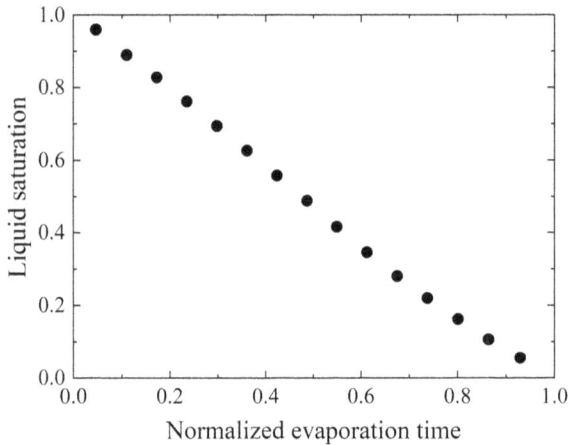

FIGURE 3.4   Variation of the liquid saturation with the normalized evaporation time. The normalized evaporation time is the evaporation time divided by the total evaporation time.

is presented in Figure 3.4. The liquid saturation varies linearly as a function of time. This indicates that the evaporation rate is almost constant. The explanation of the constant evaporation rate is as follows. As shown in Figure 3.3, the small pore throat attached to the pore body open to the environment is always filled with liquid. Combined with the long vapor transport path in the outlet tube (Figure 3.2a), this results in a quasi-constant vapor concentration in the pore body open to the environment and hence the quasi-constant evaporation rate.

For the slow evaporation in the pore network considered here, only the partially liquid filled pore with the lowest invasion threshold pressure is invaded by gas in each liquid cluster. Here, the partially liquid filled pore is the one filled with liquid and adjacent to at least one pore completely filled with gas. Gas can invade a partially

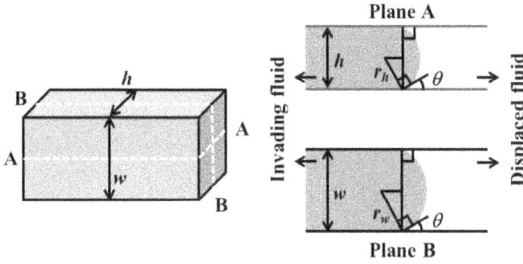

FIGURE 3.5   Schematic of a meniscus in a channel.

liquid filled pore only when the pressure difference between the gas and the liquid exceeds the invasion threshold pressure of this pore.

To understand clearly the invasion threshold pressure, we introduce here a quasi-static two-phase displacement in a pore with a rectangular cross section. The invading and displaced fluids are separated by a meniscus across which a pressure difference is established due to capillary effects:

$$P_I - P_D = \sigma \left(\frac{1}{r_w} + \frac{1}{r_h} \right) \tag{3.1}$$

where P_I and P_D are the pressure of the invading and displaced fluids, respectively; σ is the surface tension; and r_w and r_h are the radii of curvatures of the menisci in the width and height directions, respectively. These two curvature radii have direction as well as magnitude. The radius of a curvature is taken as positive if the curvature center is on the side of the invading fluid, otherwise negative. For the meniscus shown in Figure 3.5, both curvature radii are positive, and $r_h = h/2\cos\theta$ and $r_w = w/2\cos\theta$, where θ is the contact angle and h and w are the channel height and width, respectively. The contact angle is defined as the angle between the interface of the invading fluid and the displaced fluid and the interface of the displaced fluid and the solid. In the evaporation case, liquid in porous media is gradually replaced by the gas phase, similar to gas invasion; for such circumstance, gas is the invading fluid and liquid is the displaced fluid.

For the invading fluid displacing the displaced fluid, the three-phase contact line cannot move when the contact angle is larger than the advancing contact angle θ_a. In Figure 3.5, the invading fluid is the non-wetting phase. The value of the apparent contact angle is relevant to the pressure difference across the meniscus between the invading fluid and the displaced fluid, since the radius of the meniscus curvature depends on the contact angle, Eq. (3.1). As the pressure of the invading fluid in Figure 3.5 increases, the meniscus will deform, resulting in the decrease of the contact angle. Thus, the lower the contact angle is, the higher the pressure difference across the meniscus. To advance a meniscus in a pore, the pressure difference across the meniscus must be larger than a critical value so that the contact angle can be equal to or smaller than the advancing contact angle. This critical value is called the invasion threshold pressure. Obviously, the invasion threshold pressure for a pore of height h and width w is

$$P_{th} = 2\cos\theta_a \sigma \left(\frac{1}{w} + \frac{1}{h} \right) \tag{3.2}$$

Equation (3.2) can be used to determine the invasion threshold pressure for invasion into a pore throat from a pore body. But, for invasion from a pore throat into a pore body with a sudden geometrical expansion, Eq. (3.2) is not applicable. Figure 3.6 shows the process of the invading fluid entering a pore body from a pore throat. The pore throat and the pore body have the same height but different widths. The height direction is perpendicular to the plane shown in Figure 3.6. When the three-phase contact line moves from the pore throat to the interface between the pore throat and the pore body, the apparent contact angle jumps from θ_a to $\theta_a + 90°$ due to the sudden geometrical expansion, Figure 3.6a and b. The apparent contact angle is employed to define the angle between the meniscus and the solid wall in the microfluidic pore network. As a result, the three-phase contact line is pinned, and the radius of the meniscus curvature along the width direction is $r_w = w_t/2\sin\theta$, where w_t is the width of the pore throat. As the invading fluid flows, the meniscus grows and the apparent contact angle reduces from $\theta = \theta_a + 90°$ to θ_a, Figure 3.6b–d. During this period, r_w is varied, but the radius of the meniscus curvature along the height direction remains unchanged $r_h = h/2\cos\theta_a$ since the pore throat and the pore body have the same height h. Therefore, as the apparent contact angle varies from $\theta = \theta_a + 90°$ to θ_a, the pressure difference across the meniscus, determined by Eq. (3.1), is maximum at $\theta = \max(90°, \theta_a)$, i.e., $90°$ in drainage of a non-wetting fluid displacing a wetting fluid and θ_a ($>90°$) in imbibition of a wetting fluid displacing a non-wetting fluid. As the contact angle reduces to θ_a, the three-phase contact line begins to move along the wall of the pore body, Figure 3.6d and e. The radius of meniscus curvature along the width direction then increases, resulting in a lower pressure difference across the meniscus, Eq. (3.1). Hence, the invasion threshold pressure for invasion into a pore body from a pore throat with the same height h is

$$P_{th} = 2\sigma \left(\frac{\sin\left[\max(90°, \theta_a) \right]}{w_t} + \frac{\cos\theta_a}{h} \right) \tag{3.3}$$

where w_t is the width of the pore throat.

The threshold pressure to invade a pore throat is smaller than that to burst from it into a pore body, Eqs. (3.2) and (3.3). This is owing to the sudden geometrical expansion at the interface between the pore throat and the pore body. It seems that such geometrical expansion serves as a valve to control the movement of the invading fluid. This is called the capillary valve effect (CVE). During two-phase flow in a porous material, two or more pore throats adjacent to a pore body can be invaded before invasion into this pore body.

Figure 3.7 shows the process of invading fluid entering a pore body from two adjacent pore throats. Pore throats A and B are invaded and $w_A > w_B$. All the pores have

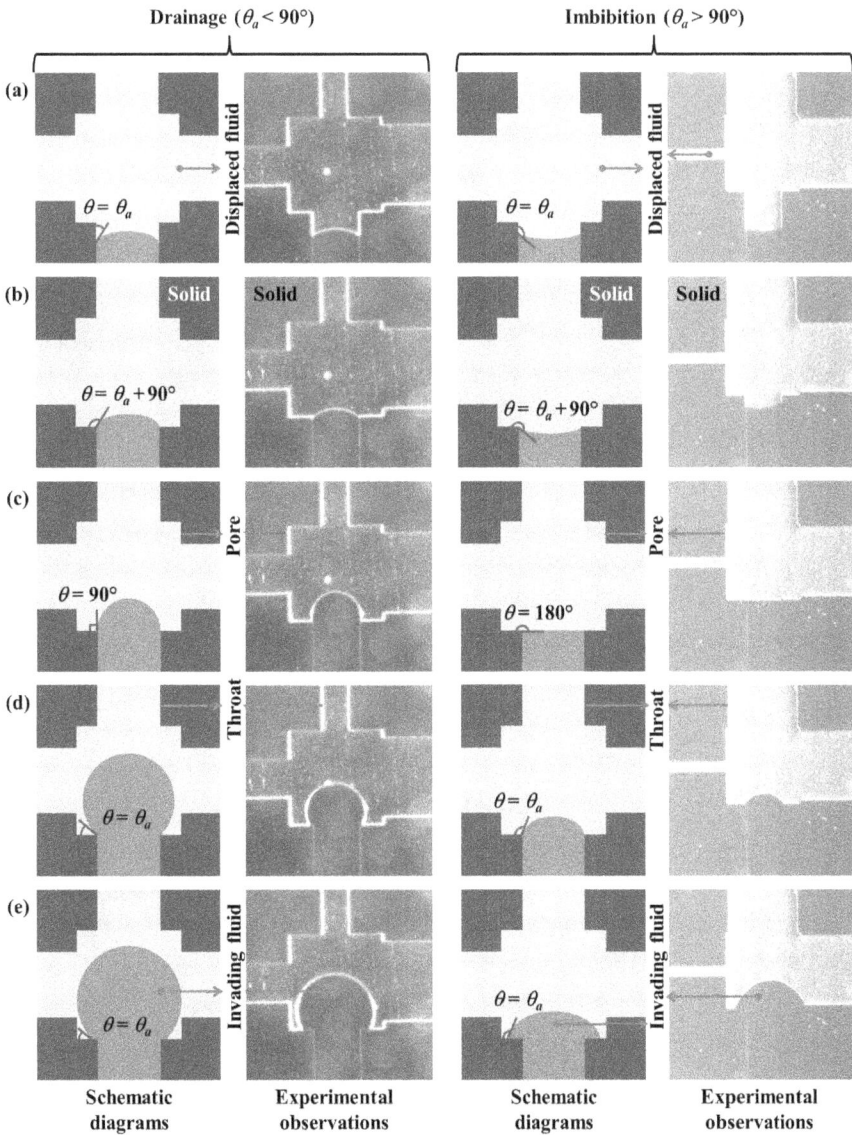

FIGURE 3.6 Schematic and experimental observation of bursting invasion into a pore body. (a) Advancement of the three-phase contact line in the pore throat. (b–d) Evolution of the meniscus when the three-phase contact line in pinned at the interface between the pore throat and the pore body. (e) Advancement of the three-phase contact line along the wall of the pore body.

the same height h. Menisci A and B have the same curvature radii since the pressure differences across them are identical when two-phase invasion is dominated by capillary forces. This also implies $\theta_A > \theta_B$. The menisci merge at $\theta_A > \max(90°, \theta_a)$. The curvature radii are $r_w = w_A/2\sin\theta_A$ before merging, Figure 3.7a and b. After merging,

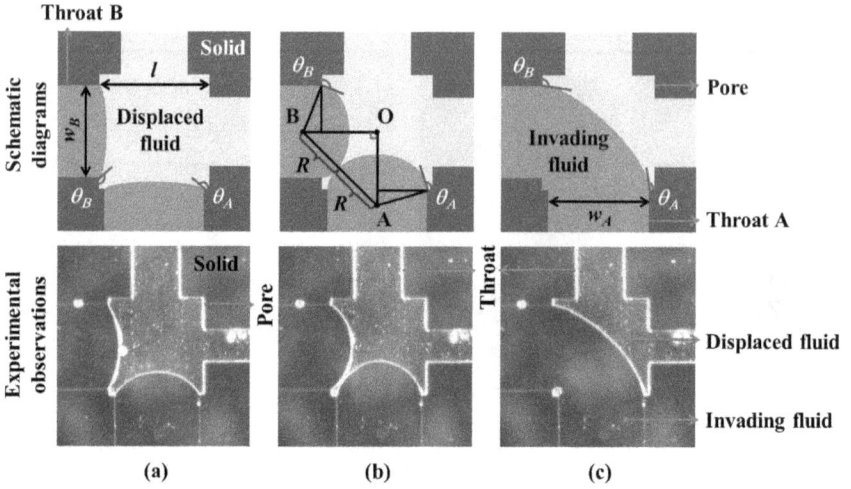

FIGURE 3.7 Schematic and experimental observation of merging invasion into a pore body: (a) before touching of two menisci, (b) at the moment of touching, and (c) after touching of two menisci.

a new meniscus forms, leading to a larger r_w, Figure 3.7c. The pressure difference across the menisci is the largest at the moment of merging, Eq. (3.1). This type of invasion into a pore body is called merging invasion (Cieplak and Robbins, 1990). The invasion threshold pressure is

$$P_{th} = \sigma \left(\frac{1}{R} + \frac{2\cos\theta_a}{h} \right) \tag{3.4}$$

where R is the radius of curvature along the width direction at the moment of menisci merging. As elucidated in Figure 3.7b, the value of R can be determined by the relationship $AB^2 = OA^2 + OB^2$:

$$(2R)^2 = \left[\frac{l}{2} + \sqrt{R^2 - \left(\frac{w_A}{2} \right)^2} \right]^2 + \left[\frac{l}{2} + \sqrt{R^2 - \left(\frac{w_B}{2} \right)^2} \right]^2 \tag{3.5}$$

At the moment of menisci merging, θ_A is larger than the value of max $(90°, \theta_a)$, which implies

$$R < \frac{w_A}{2\sin\left[\max(90°, \theta_a) \right]} \tag{3.6}$$

The menisci cannot merge at $\theta_A \geq 145°$ since the pore bodies are square and larger than their adjacent pore throats. This means

$$R > \frac{w_A}{\sqrt{2}} \tag{3.7}$$

The value of R can be determined from Eqs. (3.5)–(3.7). If no solution is found, the merging invasion shown in Figure 3.7 will not occur and the pore body will be invaded by bursting invasion from pore throat A.

3.3 EVAPORATION IN THE SILICON-GLASS BASED MICROFLUIDIC PORE NETWORK WITH CORNER LIQUID FILMS

In order to disclose the role of the corner liquid films, the evaporation in a silicon-glass based microfluidic pore network is introduced in this subsection (Wu et al., 2020). The liquid is ethanol with purity 99.99%.

3.3.1 EXPERIMENTAL SETUP

The microfluidic pore network consists of 5×5 large pore bodies connected by small pore throats; see Figure 3.8a. All the pores have the same height of 50 μm in the direction perpendicular to the plane shown in Figure 3.8a. Hence, we consider pore and meniscus structures mainly in this plane hereafter unless otherwise specified. All the pore bodies are square and have a side length of 1 mm. The distance between the centers of two adjacent pore bodies is 2 mm. All the pore throats are rectangular and have the same length of 1 mm. The widths of pore throats, w, are randomly distributed in the range of 0.14–0.94 mm. The pore network is connected to the environment through a 2 mm long and 0.5 mm wide pore throat. This pore throat is also called the outlet pore.

The cross section of a pore in the pore network is rectangular, Figure 3.8b. However, the corner is actually rounded, not sharp. A pore has four walls: a top glass wall, two side silicon walls, and one bottom silicon wall. Each solid element, surrounded by pores, has eight side walls, Figure 3.8a. The length of each side wall of a solid element can be determined by the sizes of connected pores. Note that a side wall can be connected to two corner films (Figure 3.8c): one is at the corner between the side and top walls and the other is at the corner between the side and

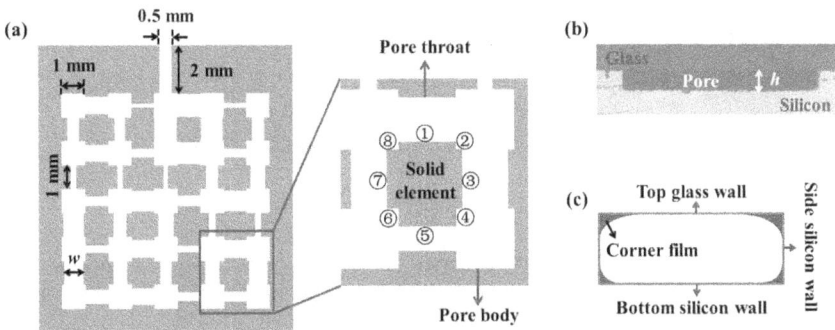

FIGURE 3.8 (a) Schematic of the microfluidic pore network used in the experiment. (b) Cross section of the outlet pore in the pore network. (c) Schematic of corner films in a pore.

bottom walls. A side wall is saturated or filled when it is covered by corner films. The contact angle is defined as the angle between the gas–liquid (ethanol) interface and the liquid–solid interface. The apparent contact angle for the silicon wall, $\theta_{si} = 35°$, is evaluated based on the shape of gas–liquid interface (meniscus) obtained in the evaporation experiment. The contact angle for the glass wall is $\theta_g = 0°$.

The pore network is initially immersed into ethanol (purity 99.99%) contained in a plastic cylinder depressurized by a vacuum pump so as to saturate the pore network with ethanol. Then, the pore network is horizontally placed in a chamber with almost constant temperature (25°C $\pm$ 2°C) and relative humidity (32% $\pm$ 2%). Time variation of the liquid distribution in the pore network is recorded by a camera (Nikon D810) equipped with a macro lens (AF-S VR Micro-Nikkor 105 mm f/2.8 IF-ED).

Since the cross section of the pore in the pore network is rectangular, it is not easy to observe the corner films directly in the pore network during evaporation. But the variation of the liquid configuration in the pore network can indicate the existence of corner films. As shown in Figure 3.9, during drying, gas invades into pore body A, transforming the meniscus curvature from concave to convex (to the gas phase), see the images from $t=0$ to 70 minutes. This is probably due to the contaminants in the pore body A. As a result, the liquid pressure increases in the pore body A, thereby

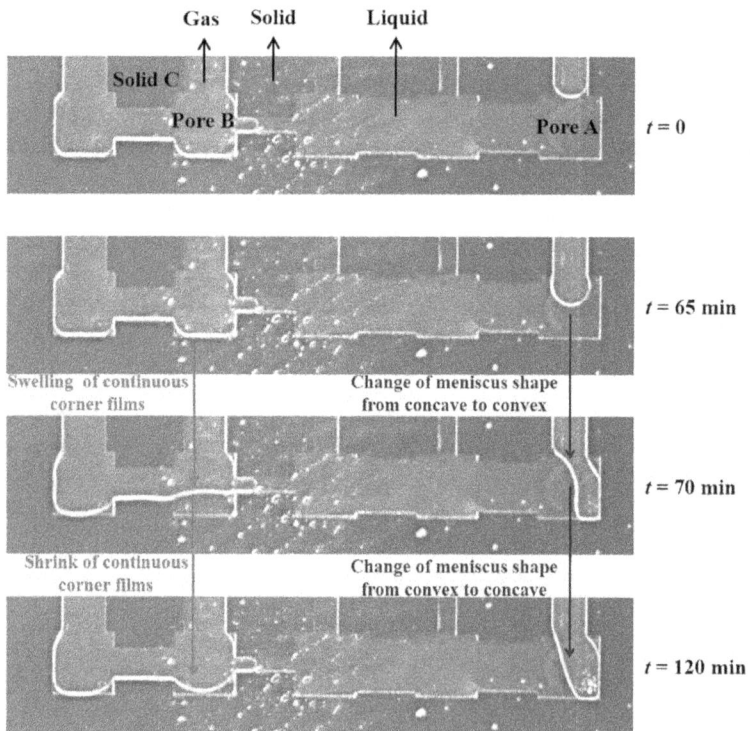

FIGURE 3.9   Swelling of continuous corner films and shrinkage of liquid during evaporation in the pore network.

resulting in higher liquid pressure in other pores, e.g., pore body B. The increase of liquid pressure leads the corner films in the pore body B, if any, to swell. This swelling of corner films explains well why the pore body B is filled by liquid during the gas invasion into pore body A, see image of $t = 70$ minutes in Figure 3.9. As the meniscus in pore body A becomes concave again, the pressure of liquid is decreased, resulting in the shrinkage of liquid in the pore body B, see image of $t = 120$ minutes in Figure 3.9. The variation of the liquid configuration in pore body B demonstrates that there are continuous corner films connecting the pore body B and the bulk liquid in the pore network shown in Figure 3.9.

Moreover, menisci are white and bright in the experimental images, Figure 3.9. Some walls attached to empty pores (e.g., pore body B) show similar color as menisci, whereas some others (e.g., the walls of solid element C) do not. By comparing the colors of walls and menisci, we can also infer that there exist corner films in the pore network.

3.3.2 EXPERIMENTAL RESULTS

The evolution of liquid distribution in the pore network during drying is shown in Figure 3.10. In this figure, the gas, liquid, and solid are shown in light gray, light black, and dark gray, respectively. It is not easy to observe the corner films directly in the experimental images, although the gas–liquid interfaces are white and bright. In the experiment, two small gas bubbles exist at the bottom-left and bottom-right of the pore network. These two gas bubbles remain static during drying, Figure 3.10; hence, their influences are negligible. After $S_t = 0.52$, gas invades the second pore body from the left at the bottom of the pore network. As a result of this invasion, the meniscus

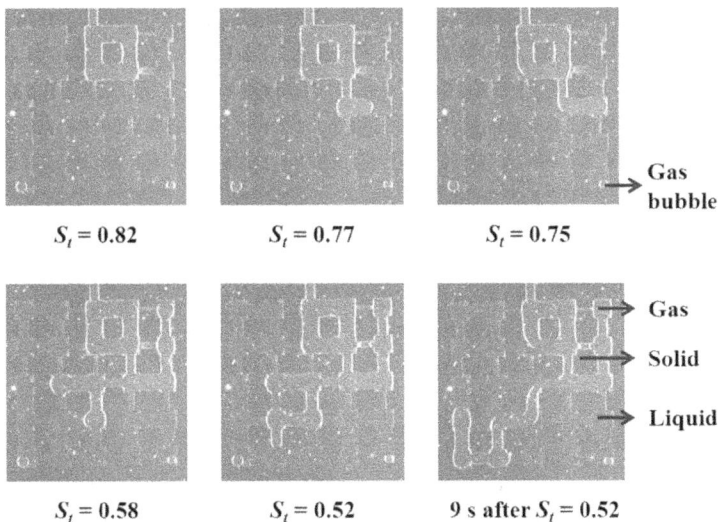

$S_t = 0.82$ $S_t = 0.77$ $S_t = 0.75$

Gas bubble

$S_t = 0.58$ $S_t = 0.52$ 9 s after $S_t = 0.52$

Gas

Solid

Liquid

FIGURE 3.10 Variation of the liquid distribution in the pore network during evaporation.

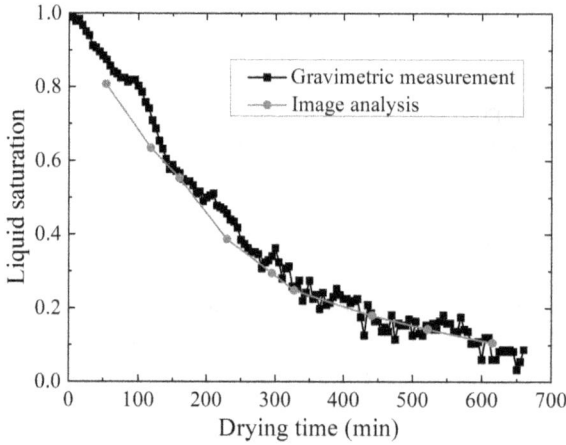

FIGURE 3.11 Variation of the total liquid saturation, S_t, with the evaporation time, t, gained by two experimental methods: image analysis and weight measurement.

in the pore body becomes unstable and enters the pores at the lower-left zone of the pore network quickly; at the same time, some empty pores at the center of the pore network are refilled by liquid. This unstable phenomenon is explained in detail in Section 3.4. The distribution of the liquid after $S_t = 0.52$ is not shown in Figure 3.10.

The evaporation rate curve deduced from the experiment is shown in Figure 3.11. Two experimental methods, i.e., image analysis and gravimetric measurement, are employed to obtain the drying rate curve. As can be seen, the results obtained from these two methods are rather similar. The evaporation rate curve can be divided into two stages: a relative high evaporation rate stage at $S_t > 0.2$ and a lower evaporation rate stage at $S_t < 0.2$. The evaporation rate at $S_t > 0.2$ is almost constant. We recall that for the evaporation in the PDMS-based pore network introduced in Section 3.2, the constant evaporation rate is due to the filled pore throat attached to the pore open to the environment via a long channel. But for evaporation in the silicon-glass based pore network in this subsection, the drying rate decreases as the total liquid saturation varies from $S_t > 0.2$ to smaller $S_t < 0.2$, although the pore throats attached to the pore body open to the environment are always filled with liquid. Since the corner liquid films exist in the pore network, the reduction of the drying rate at $S_t < 0.2$ indicates that the corner films at the outlet pore are removed.

To understand the change of the evaporation rate observed in the experiment, we present in Figure 3.12 the liquid distribution in the zone near the outlet pore from the evaporation time $t = 322.7$ to 351.2 minutes. These two times correspond to the total liquid saturation $S_t = 0.25$ and 0.18, which are in the transition region between the high and low evaporation rate stages.

As shown in Figure 3.12, as gas invades pore body F, residual liquid can be clearly observed in pore body I adjacent to the outlet pore. The side walls in the outlet pore are white and bright, indicating the existence of corner films. As gas invades pore throat G, pore body I is connected to liquid in pore throat G through corner films, increasing the resistance for liquid flow from moving menisci to corner films in the

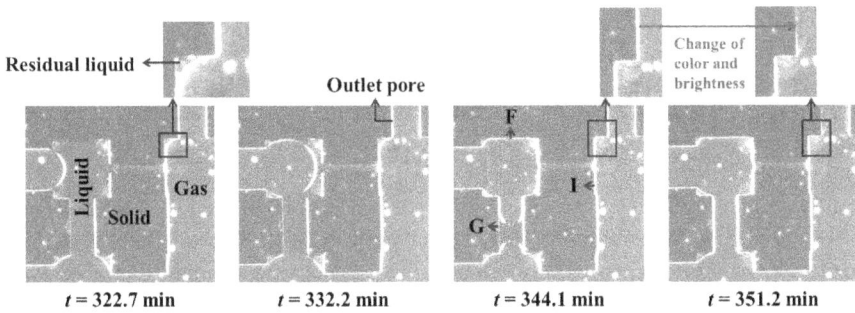

FIGURE 3.12   Gas invasion process in the upper left zone of the pore network during evaporation.

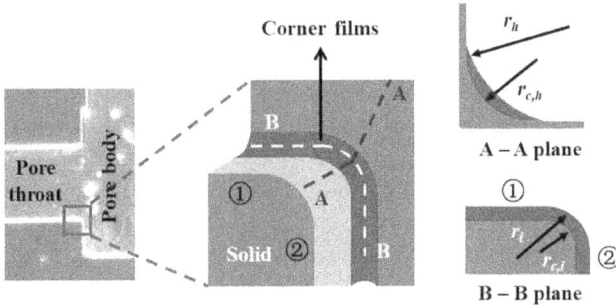

FIGURE 3.13   Schematic of the morphology of the meniscus at the intersection between two side walls.

outlet pore. As a result, liquid pressures in the corner films in the outlet pore and in pore body I are reduced, and the residual liquid in pore body I is rather small (image of $t = 344.1$ minutes in Figure 3.12).

After pore throat G is emptied (image of $t = 351.2$ minutes in Figure 3.9), the resistance is rather large for liquid flow from the moving menisci at the lower-right zone of the pore network to the corner films at the outlet pore. Hence, liquid pressure in the corner films in outlet pore and in filled pore throat H is rather small so as to pump enough liquid to compensate the evaporative loss at the outlet pore. Since pores have smaller invasion threshold pressures than corners, it is expected that pore throat H will be invaded before the corner films in the outlet pore. In the experiment, however, pore throat H is still filled; whereas the change of color and brightness of the side wall in the outlet pore indicates that the corner films therein are removed, see the right two images in Figure 3.12. In brief, we believe that the corner films removal in the outlet pore is the main reason for the two drying stages observed in the experiment.

The removal of the corner films in the outlet pore is attributed to an effect referred to as the "capillary scissors". This effect is induced by the local convex topology of the solid matrix. We show in Figure 3.13 schematically the corner films at intersection between side wall 1 in a pore throat and side wall 2 in a pore body. The intersection of these two side walls is not sharp but rounded. In the B–B plane shown in Figure 3.13, the solid element is convex toward the fluid at the intersection of side walls 1 and 2. The pressure difference across the meniscus at this intersection is $P_g - P_l = \sigma(1/r_h - 1/r_i)$. Here, r_h and r_i are the curvature radii shown in Figure 3.13, both of which are positive.

If we assume that the liquid pressure in the corner film at the intersection point is the same as that in the neighboring side walls, e.g., side walls 1 and 2 in Figure 3.13, then r_h at the intersection point is smaller, since $1/r_i$ is zero for menisci at side walls. As the liquid pressure decreases, r_h reduces. When r_h reduces to the curvature radius, $r_{c,h}$, of the intersection between the side wall and the top (or bottom) wall, the corner films at side walls 1 and 2 are separated; at this moment, r_i is equal to the curvature radius of the intersection between side walls 1 and 2, $r_{c,i}$. Hence, the threshold pressure for gas to invade the corner film at the convex intersection between two side walls is $\sigma(1/r_{c,h} - 1/r_{c,i})$, which decreases with smaller $r_{c,i}$, and hence, could be lower than that of a pore.

The invasion threshold pressure for gas to invade the corner film at a side wall is $\sigma/r_{c,h}$, which is larger than those of pores since $r_{c,h}$ is smaller than pore sizes. To this end, removal of the corner film observed in Figure 3.12 could be attributed to the fact that gas invades the corner film at the convex intersection between two side walls (its threshold pressure is smaller than those at side walls), resulting in the interruption of the corner film along the side walls. Such convex intersection at the solid element seems to serve as scissors to cut off the corner films. Therefore, we call this effect as the capillary scissors.

3.4   CAPILLARY INSTABILITY DURING EVAPORATION IN THE SILICON-GLASS BASED MICROFLUIDIC PORE NETWORK

As we mentioned in Figure 3.10, during evaporation in the silicon-glass based micro-fluidic pore network, liquid can refill the pores occupied by gas. We find that such refilling of liquid into pores occupied by gas can result in the formation of a gas bubble in the pore network; the gas bubble, once formed, moves rapidly until a stable configuration is reached; see Figure 3.14. The time for the liquid refilling and the bubble movement is about 30 s, which is much smaller than that associated with the whole drying process (about 30 hours). Despite such a small time scale, the liquid refilling into the pores and the bubble movement, without a doubt, influence the liquid distribution in the pore network and consequently the drying kinetics. As we will show later, the refilling of pores by liquid and the bubble movement shown in Figure 3.14 are attributed to the capillary instability induced by the heterogeneity in wettability and structure of the pore network. It should be noted that for the micro-fluidic pore network used in the experiments, the pore thickness (50 μm) is much smaller (by a factor up to almost 20) than the pore widths. As a result, a small change in the thickness and the wettability has a much stronger impact in this particular system. In this subsection, the capillary instability that induced the gas–liquid displacement in the pore network during evaporation is thoroughly investigated.

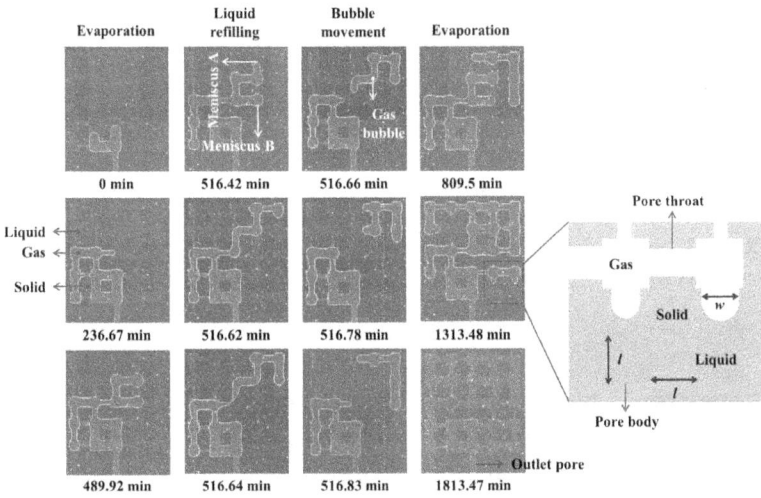

FIGURE 3.14 Refilling of pores with liquid and bubble formation and movement observed during evaporation of the pore network composed of pore bodies and pore throats (the pore length and width are also illustrated). The left and right columns are the evaporation process. The middle two columns are capillary instability induced liquid refilling and bubble formation and movement processes.

3.4.1 Experimental Setup

To disclose the capillary instability induced the gas–liquid displacement in the pore network, it is needed to know the variation of the speed of the moving meniscus and the curvature of the moving meniscus. In addition, the capillary instability is due to the heterogeneity in wettability. Thus, we also need to know the contact angle of each pore in the pore network. The following image processing procedures are employed to determine the speed of the moving meniscus, the curvature of the moving meniscus, and the contact angle of each pore.

The image processing procedure to calculate the gas phase area and the average meniscus velocity is illustrated in Figure 3.15 and is detailed as follows:

(1) Figure 3.15b is obtained by subtracting the background image from the raw image in Figure 3.15a.
(2) Figure 3.15c is obtained by applying the mean filter to Figure 3.15b.
(3) Binarization operation is applied to image in Figure 3.15c from which gas–liquid interface is extracted; see Figure 3.15d.
(4) Each pixel in Figure 3.15d is checked if it is located between two white pixels, i.e., inside the gas phase; if it is, then it is set to be white (Figure 3.15e).
(5) The area of the gas-filled region is equal to the number of pixels in the white region in Figure 3.15e multiplied by the actual area of each pixel. The average velocity of a moving meniscus in a pore is defined as the rate of the change of the gas (or liquid) area in this pore divided by the length of the pore body, which is constant.

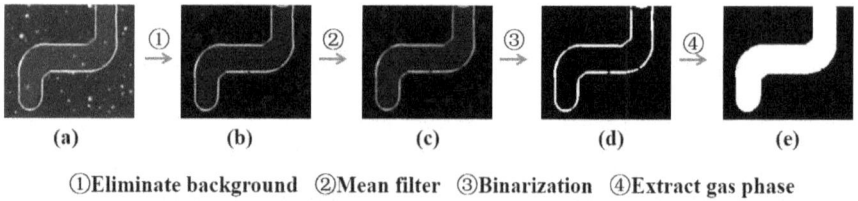

(a)   (b)   (c)   (d)   (e)

①Eliminate background ②Mean filter ③Binarization ④Extract gas phase

FIGURE 3.15 Image processing procedure to calculate the gas phase area and the average meniscus velocity.

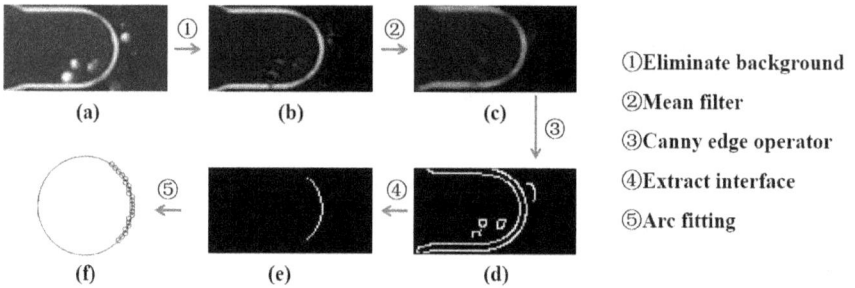

①Eliminate background

②Mean filter

③Canny edge operator

④Extract interface

⑤Arc fitting

FIGURE 3.16 Image processing procedure to determine the curvature radius of the meniscus.

The image processing procedure applied to determine the curvature radius of the meniscus is illustrated in Figure 3.16 and detailed as follows:

(1) Figure 3.16b is obtained by subtracting the background image from the raw image in Figure 3.16a.
(2) Figure 3.16c is obtained by applying the mean filter to Figure 3.16b.
(3) Figure 3.16d is obtained by using the Canny edge operator (Canny, 1986) to extract all edging points at the gas–liquid interfaces. Canny edge operator is used here because it can extract continuous edges, whereas the edges obtained by the binarization operation, as shown in Figure 3.15d, can be discontinuous since the images in Figure 3.15 and this figure are obtained by Nikon camera, and their resolutions are not high enough.
(4) Figure 3.16e is obtained by extracting the right edge in Figure 3.16d. Two edges are extracted by the Canny edge operation; see Figure 3.16d. Since the menisci in the pores are convex toward liquid, the left edge in Figure 3.16d can be the edge of the meniscus at the top or bottom wall. The right edge can be the edge of the meniscus in the plane (perpendicular to the height direction) located at the center of the pore. Here, we use the right edge in Figure 3.16d to evaluate the curvature radius of the meniscus. For this reason, the right edge in Figure 3.16d is extracted; see Figure 3.16e.

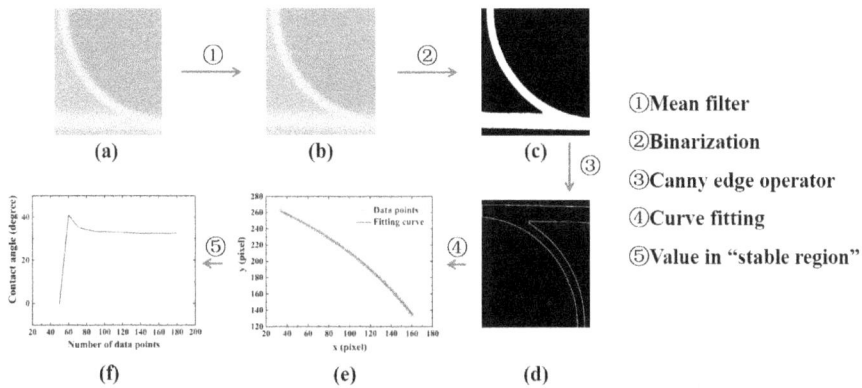

FIGURE 3.17 Image processing procedure to gain the contact angle between the gas–liquid interface and the gas–solid interface.

(5) Figure 3.16f is obtained by applying the arc fitting approach to the edge in Figure 3.16e. It is reasonable to assume that the menisci in the microfluidic pore network are circular. The arc fitting method developed by Tao et al. (2018) is applied to fit the edge shown in Figure 3.16e, and the fitted circle is shown in Figure 3.16f. As compared to the studies (Gander et al., 1984; Chernov et al., 2005), the fitting method (Tao et al., 2018) is more accurate for fitting short arc. The radius of the fitted circle is the curvature radius of the meniscus in the plane perpendicular to the pore height direction. The details of the arc fitting method can be found in Tao et al. (2018) and are not repeated here.

The image processing procedure to gain the apparent contact angle between the gas–liquid interface and the gas–solid interface is illustrated in Figure 3.17 and detailed as follows:

(1) Figure 3.17b is obtained by applying the mean filter to the gray-scale image in Figure 3.17a. The background removal is not used here. If the background removal is performed, the triple point where the liquid, gas, and solid meet cannot be found. The value of the contact angle cannot be obtained without the information of the triple point.

(2) Figure 3.17c is obtained by using the binarization operator to Figure 3.17b to extract the gas–liquid interface and the solid surface. Here, we use the binarization operator to get the gas–liquid interface; the reason is that the images are gained from microscope and have a high resolution. The gas–liquid interface and the solid surface are whiter than the gas, liquid, and solid zones, and hence can be easily discerned by the binarization operator. But for the images gained from the Nikon camera, the gas–liquid interface and the solid surface cannot be captured by the binarization operator because of the low resolution of these images.

(3) Figure 3.17d is obtained by using the Canny edge operator (Canny, 1986) to extract all edge points at the gas–liquid interface in Figure 3.17c after it is rotated so that so that the two straight edges are horizontal and the curved edges are convex to right.

(4) Figure 3.17e is obtained by using the cubic polynomial function that is used to fit the curve near the triple point for the right gas–liquid edge shown in Figure 3.17d. The detailed fitting procedures can be found in Atefi et al. (2013). The contact angle is obtained straightforwardly based on the slope of the fitted curve at the triple point.

(5) Figure 3.17f shows the variation of the contact angle obtained by the fitting method in step (4) with the number of data points used in fitting. As the number of data points increases, e.g. to more than 130, the contact angle reaches a stable region. The apparent contact angle in such stable region is hence used to characterize the surface wettability.

The image processing procedures for the contact angle are applicable only to the high-resolution images obtained from the microscope. The image processing procedures introduced in Figures 3.15 and 3.17 can be applied to the all images obtained in our experiments.

The advancing and receding menisci are defined as those moving toward the liquid and gas phases, respectively. The speed of a moving meniscus in a pore is defined as the rate of liquid flow in this pore divided by the cross-sectional area of the pore body (all pore bodies have the same cross-sectional area). Since the pore network structure is etched in silicon wafer in our microfluidic model, we can get the contact angle (the angle between the gas–liquid interface and the liquid–solid interface) for the silicon surface, θ_s, based on the image analysis. The contact angle hysteresis for the silicon surface is small and can be neglected. The contact angle for the glass surface is $\theta_g = 0°$. Here, we find from the experiment that the triple line, where gas, liquid and solid meet, at the side wall of the pore moves toward the liquid phase when the angle between the gas–liquid interface and the liquid–solid interface is smaller than θ_s, and moves to the gas phase when this angle is larger than θ_s.

3.4.2 EXPERIMENTAL RESULTS

The detailed menisci movement during refilling of pores shown in Figure 3.14 (the second column from the left) is presented in Figure 3.18. The images in Figure 3.14 are obtained by the camera (Nikon D810), which has a larger view field but a lower resolution as compared to the microscope (Olympus IX73 with a 4X lens). To understand in detail the menisci displacement during the liquid refilling process, we repeat the evaporation experiment by using the microscope to get high resolution images of gas–liquid menisci in the pore network so as to get the accurate curvature radii and contact angles of menisci. The microscope cannot capture the whole zone of the pore network. Only the zone of 3×3 pore bodies at the upright of the pore network is shown in Figure 3.18. The location of the zone captured by the microscope is marked by the red box in Figure 3.14. Menisci A and B shown in Figure 3.18 are also

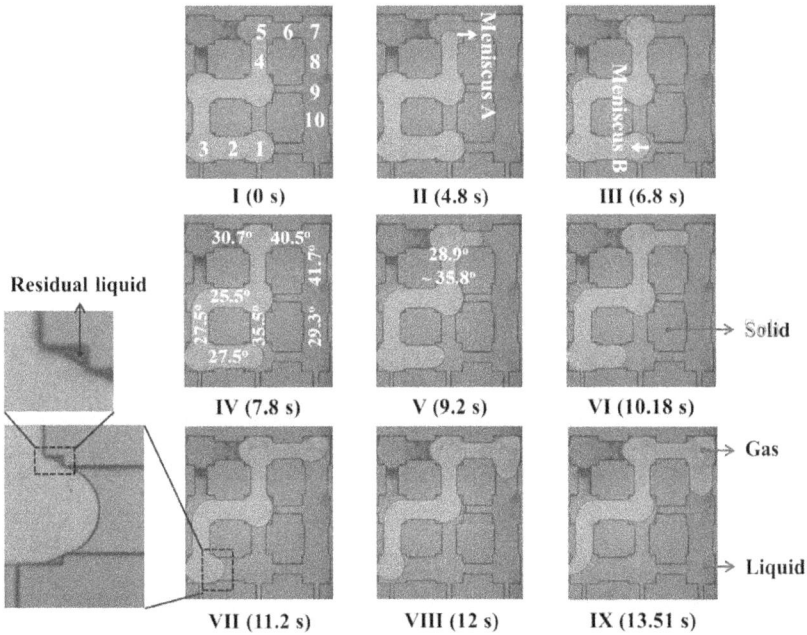

FIGURE 3.18 Evolution of the moving menisci during capillary instability induced refilling of pores in the top right zone of the pore network (i.e., the zone marked by the red box in Figure 2.14). Menisci A and B are moving, while the other menisci remain almost static. The pores that can be invaded by menisci are number. The contact angles of the pore throat are also shown in the images of IV and V.

illustrated in Figure 3.14. The movement of menisci A and B in Figure 3.18 is similar to that in Figure 3.14, indicating the repeatability of the experiments.

Menisci A and B shown in Figure 3.18 are the advancing and the receding menisci, respectively. The variations of speeds and curvature radii of these two menisci are depicted in Figure 3.19. In Figure 3.19b, r_i is the curvature radius of the meniscus in the plane shown in Figure 3.18. The value of r_i is obtained from the image analysis. The curvature radius in the plane perpendicular to the one shown in Figure 3.18 is denoted as r_h. At stage I in Figure 3.18, the meniscus A is at the entrance of pore body 5. At the interface between a pore throat and a pore body, a sudden geometrical expansion exists, which can hinder the meniscus movement, i.e., the so-called CVE. More detailed explanation on the CVE can be found in Wu et al. (2016, 2017). The evolution of meniscus A during its invasion into pore body 5 is similar to that shown in Figure 3.6. The apparent contact angle of meniscus A at the side wall increases suddenly from θ_s to $\theta_s + 90°$ when the triple line at the side wall of pore throat 4 moves to the entrance of pore body 5. The triple line of meniscus A at the side wall is then pinned, since it cannot move toward the liquid phase until the contact angle is reduced to θ_s. However, pore throats and pore bodies have the same height; hence, the triple line at the top glass wall and the bottom silicon wall continue moving. As a result, meniscus A reshapes, and thus, the apparent contact angle of meniscus A at

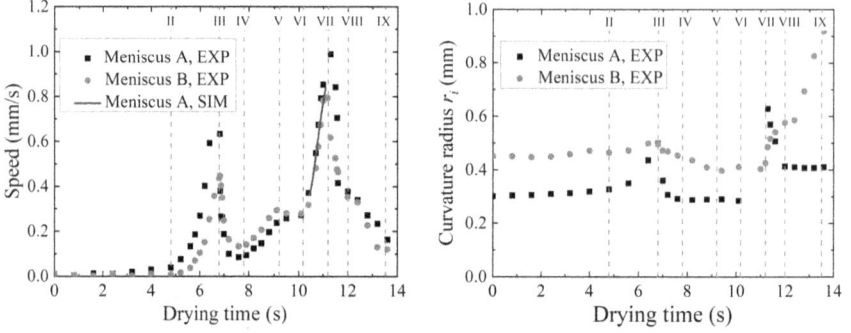

FIGURE 3.19   Variation of (a) menisci speeds and (b) curvature radii r_i during refilling of pores shown in Figure 3.18. Menisci A and B as well as various evaporation stages are illustrated in Figure 3.2.

the side wall reduces. As this apparent contact angle decreases from $\theta_s + 90°$ to $90°$, the curvature radius, r_i, of meniscus A decreases. But, as the contact angle further reduces from $90°$ to θ_s, the curvature radius, r_i, increases, resulting in a lower capillary pressure, P_c. The capillary pressure is defined as $P_c = P_g - P_l = \sigma(1/r_i + 1/r_h)$, where σ is the surface tension, P_l the liquid pressure, and P_g the gas pressure. Since P_g is constant during evaporation, the lower the capillary pressure is, the higher the liquid pressure is. The increase of liquid pressure at meniscus A results in higher liquid pressure in the pore network. The increased liquid pressure can lead some other menisci in the pore network to be unstable, e.g. meniscus B in Figure 3.18. A meniscus is unstable when $P_l - P_g$ across the meniscus is larger than the invasion threshold pressure for the meniscus to refill the pore occupied by the gas phase. A meniscus cannot invade a pore when the pressure difference in the invading and displaced phases across this meniscus is smaller than the invasion threshold pressure of the pore. The instability of meniscus B is induced by the variation of the capillary force of the advancing meniscus A. We call this the capillary instability. This capillary instability effect is the reason for the pore refilling and the bubble formation and displacement t is shown in Figure 3.14.

The displacement of menisci A and B in Figure 3.18 depends on the difference between the liquid pressures at these two menisci, $P_{l,A} - P_{l,B}$. Based on the definition of the capillary pressure, the value of $P_{l,A} - P_{l,B}$ is determined as

$$P_{l,A} - P_{l,B} = \sigma\left(\frac{1}{r_{i,B}} - \frac{1}{r_{i,A}}\right) + \sigma\left(\frac{1}{r_{h,B}} - \frac{1}{r_{h,A}}\right) \tag{3.8}$$

Equation (3.8) indicates that the displacement of menisci depends on the curvature radii, r_i and r_h. For instance, from stages II to III shown in Figure 3.18, the speeds of the moving menisci are always increasing as meniscus A enters pore body 5, Figure 3.19a. The higher speed is due to the increased $r_{i,A}$ (Figure 3.19b), which in turn results in the increased $P_{l,A} - P_{l,B}$, Eq. (3.1). However, as shown in Figure 3.19b, $r_{i,A} < r_{i,B}$ from stages I to III. If menisci A and B have the same value of r_h, then

$P_{l,A} - P_{l,B} < 0$, Eq. (3.1); and the capillary instability induced gas–liquid displacement shown in Figure 3.1 will not occur. From this point of view, it can be inferred that $r_{h,A} > r_{h,B}$.

During stages I to III shown in Figure 3.18, both menisci A and B are in the pore bodies, and the triple lines, where gas, liquid, and solid meet, at the side walls are pinned at the entrance of the pore bodies and are not moving. Hence, no corner liquid films form when these two menisci advance in the pore bodies. For this reason, r_h can be expressed as $r_h = h/(1 + \cos\theta_s)$, and the effects of the corner liquid films on the capillary pressure, e.g. Wong et al. (1992), are not considered. As a result, to get $r_{h,A} > r_{h,B}$, the contact angle, θ_s, of pore body 5 should be larger than that of pore body 1. Nevertheless, it is not easy to obtain the values of θ_s in the pore bodies, since we cannot observe the moving triple line along the side walls of pore bodies. By contrast, we can get the values of θ_s in the pore throats. As shown in Figure 3.18 (image at stage IV), pore throats have different values of θ_s.

For a meniscus in a pore throat, e.g., meniscus A in pore throat 6 and meniscus B in pore throat 2 shown in image V in Figure 3.18, the capillary pressure for this meniscus shall be affected by the corner liquid films, since the lengths of the corner liquid films vary as the meniscus moves. As a result, r_h in Eq. (3.8) cannot be expressed simply as $r_h = h/(1 + \cos\theta_s)$. To calculate the capillary pressure for a meniscus in a pore with corner liquid films, two methods can be exploited. The first is the so-called Mayer and Stowe-Prince (MS-P) method (Mayer and Stone, 1965; Princen, 1969). In the second method proposed by Wong et al. (1992), the shape of the meniscus in the pore is determined from which the capillary pressure is obtained. The results obtained by these two methods are almost the same. The MS-P method is more convenient for the calculation, and therefore, is employed to determine the capillary pressure. To this end, the capillary pressure, $P_c = P_g - P_l$, for a moving meniscus in a pore throat i is determined as

$$P_c = \frac{\sigma(k_1 + 2k_2k_4)}{w_i h - 2k_3 k_4^2} \tag{3.9a}$$

$$k_1 = (2h + w_i)\cos\theta_{s,i} + w_i \tag{3.9b}$$

$$k_2 = \pi - 3\theta_{s,i} - (2\cos\theta_{s,i} - 3\sin\theta_{s,i} + 2)\cos\theta_{s,i}, \tag{3.9c}$$

$$k_3 = \cos^2\theta_{s,i} + \cos\theta_{s,i} - \frac{3\sin\theta_{s,i}\cos\theta_{s,i}}{2} - \frac{\pi}{2} + \frac{3\theta_{s,i}}{2} \tag{3.9d}$$

$$k_4 = \frac{-2k_1k_3 + \sqrt{4k_1^2k_3^2 - 8k_2^2k_3wh}}{4k_2k_3} \tag{3.9e}$$

If pore throats 2 and 6 have the same contact angle, then the capillary pressure determined by Eq. (3.9) is larger for meniscus A in pore throat 6 than for meniscus B in pore throat 2. For instance, if the contact angle is 0°, the capillary pressure is 947.1 Pa for meniscus A and 926.9 Pa for meniscus B. This indicates that the liquid pressure at meniscus B is larger than that at meniscus A, and the capillary stability induced gas–liquid displacement cannot occur. By contrast, if the measured apparent contact angle (see Figure 3.18) is used for pore throats 2 and 6, then the capillary pressure for meniscus A in pore throat 6 (834.5 Pa) is lower than that for meniscus B in pore throat 2 (876.2 Pa), which implies that meniscus A advances and meniscus B recedes, consistent with the experimental observation.

The capillary instability induced two-phase displacement observed in the present study could be attributed to the wettability heterogeneity of the pore network. The wettability of the pore surface can be essentially reflected by the measured apparent contact angle. Since r_h in Eq. (3.1) cannot be expressed simply as $r_h = h/(1 + \cos\theta_s)$ when the meniscus is in the pore throat (because of the corner liquid films), we focus on the measured r_i in the following analysis so as to illustrate the variation of the meniscus shape during the capillary instability induced gas–liquid displacement shown in Figure 3.18. At stage III shown in Figure 3.18, meniscus A touches the side wall of pore body 5 as well as the entrance of pore throat 6. After this, meniscus B continues to move in pore body 1, and meniscus A enters pore throat 6, stage IV of Figure 3.18. From stages III to IV, $r_{i,A}$ reduces significantly as meniscus A enters pore throat 6 from pore body 5, whereas the decrease of $r_{i,B}$ is smaller, Figure 3.19b. As a result, $P_{l,A} - P_{l,B}$ decreases, Eq. (3.8), leading to the reduced menisci moving speed, Figure 3.19a.

After meniscus A invades pore throat 6, meniscus B continues invading the gas phase in pore body 1 until it enters pore throat 2, stage V in Figure 3.18a. During this process, $r_{i,A}$ remains constant and $r_{i,B}$ reduces, Figure 3.19b, thereby leading to increase in the pressure difference $P_{l,A} - P_{l,B}$, Eq. (2.8). Hence, the speeds of the moving menisci become higher, Figure 3.19a. From stages V to VI, both menisci A and B are in pore throats. Therefore, variations of curvature radii and speeds of moving menisci are trivial, Figure 3.19.

After stage VI, meniscus A invades pore body 7 from pore throat 6; see stage VII in Figure 3.18. During this process, $r_{i,A}$ increases due to the CVE; the value of $r_{i,A}$ is not shown in Figure 3.19b, since the shape of meniscus A, as shown in stage VII of Figure 3.18, is not circle due to the contaminant in pore body 7. Meniscus B is always in pore throat 2 from stages VI to VII, and $r_{i,B}$ is almost constant, Figure 3.19b. As a result, $P_{l,A} - P_{l,B}$ increases, Eq. (3.8), leading to higher speeds of moving menisci, Figure 3.19a.

After stage VII, meniscus A invades pore throat 8 and meniscus B enters pore body 3, stage VIII in Figure 3.18. During this process, $r_{i,A}$ decreases and $r_{i,B}$ increases, resulting in the decreased $P_{l,A} - P_{l,B}$, Eq. (3.8), and hence, the speeds of the moving menisci reduce, Figure 3.19. After the stage VIII shown in Figure 3.18, meniscus B connects two gas-filled pore throats connected to pore body 3 and $r_{i,B}$ increases as the liquid saturation in the pore body increases. By contrast, $r_{i,A}$ is almost constant, Figure 3.19b. Although $r_{i,A} < r_{i,B}$ at stage VIII, meniscus B continues refilling of pore body 3; see stage IX in Figure 3.18. When pore body 3 is completely filled by liquid, a gas bubble is formed in the pore network, Figure 3.14.

When meniscus B is entering pore body 3 from pore throat 2 (from stages VII to VIII shown in Figure 3.18), we find that the influence of the CVE induced by the sudden geometrical expansion between the pore throat and the pore body is trivial.

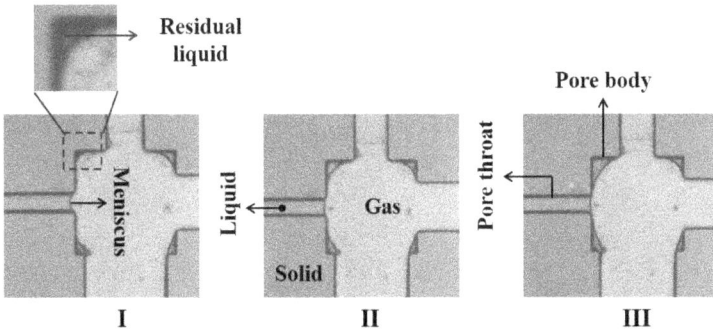

FIGURE 3.20   Growth of the residual liquid in an empty pore body due to the presence of the corner liquid films during the invasion of a meniscus into this pore body.

The CVE is suppressed by the residual liquid in pore body 3. Before meniscus B invades pore body 3, the intersectional points between pore throat 2 and pore body 3 are attached to the residual liquid. As the triple lines (at the side walls) of meniscus B move to these intersectional points, meniscus B will merge with the residual liquid to form a new meniscus. During this process, meniscus B keeps concave toward the gas phase.

For the capillary instability induced gas–liquid displacement shown in Figure 3.18, menisci A and B interact with each other through the liquid filled pores. The menisci displacement in this case corresponds to the one induced by the capillary instability with pore flow. We also observe in the experiment that the two menisci interact with each other through the corner liquid films, i.e., the menisci movement induced by the capillary instability with corner flow; see Figure 3.20. The speed of the moving menisci due to the capillary instability with the corner flow is much smaller than that with the pore flow. It has been revealed that the speed of the capillary instability induced advancing meniscus depends on the number of interacting receding menisci (Armstrong and Berg, 2013). Here, we reveal that the speed of moving menisci induced by the capillary instability also depends on the flow pattern between the interacting menisci.

The corner liquid films also can influence the invasion of a meniscus from a pore throat to a pore body. During the meniscus invasion into pore body 3 from pore throat 2 shown in Figure 3.18, the residual liquid in the pore body is always connected to the mouth of the pore throat. We also observe in the experiment that when a meniscus in a pore throat reaches the intersectional points between this pore throat and the adjacent pore body, the residual liquid in the pore body can be not attached to the mouth of the pore throat; see Figure 3.20. Owing to the presence of corner liquid films, the meniscus in the pore throat and the residual liquid in the pore body are actually connected and can interact with each other. During invasion of the meniscus from the pore throat to the pore body, the liquid pressure at the meniscus will increase due to the CVE, which in turn results in the growth of the residual liquid because of the presence of the corner liquid films. When the residual liquid grows to touch the mouth of the pore throat, the meniscus is still concave toward the

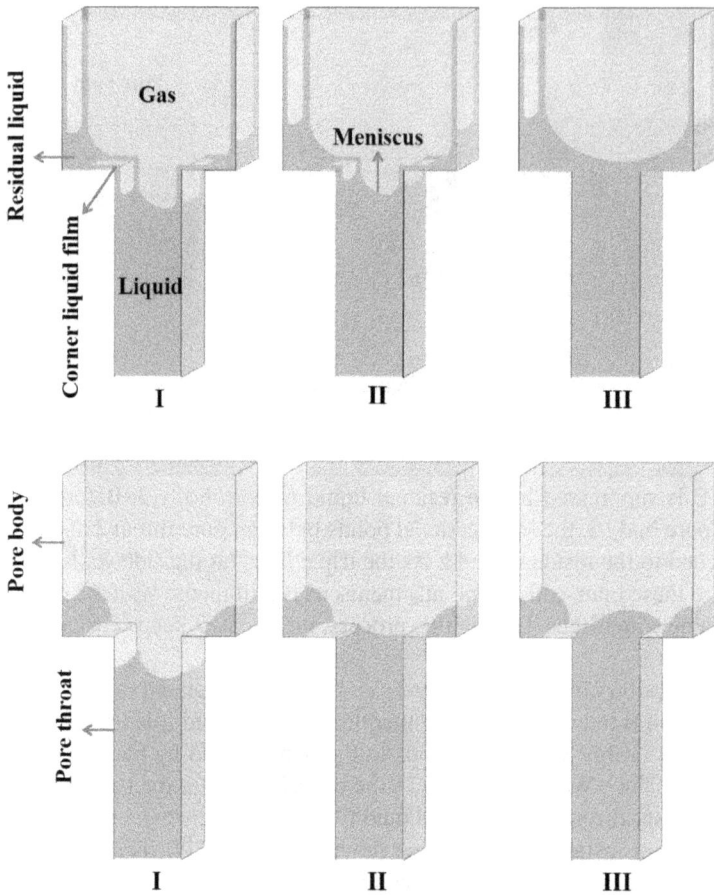

FIGURE 3.21 Schematic of a meniscus displacement from a liquid filled pore throat into a pore body in which corner liquid films are (a) present and (b) absent.

gas phase. Then, the meniscus will keep concave as it enters the pore body, similar to the invasion process shown in Figure 3.18.

To understand the role of the corner liquid films in the meniscus invasion into a pore body from a pore throat, we compare schematically the invasion processes in the cases with and without corner liquid films in Figure 3.21. In the case with the corner liquid films in the pores and the residual liquid in the pore body, the meniscus keeps concave toward the gas phase as it invades the pore body from the pore throat, as we discussed above. By contrast, in the case without the corner liquid films, the residual liquid in the pore body (if exists) is not connected to the invading meniscus. To this end, the meniscus has to change from concave to convex (toward the gas phase) during its invasion into the pore body, attributed to the CVE. Hence, the invasion threshold pressure for the meniscus to invade the pore body is larger in the case without the corner liquid films than in the case with the corner liquid films. If there were no corner liquid films in the pores and no residual liquid in the

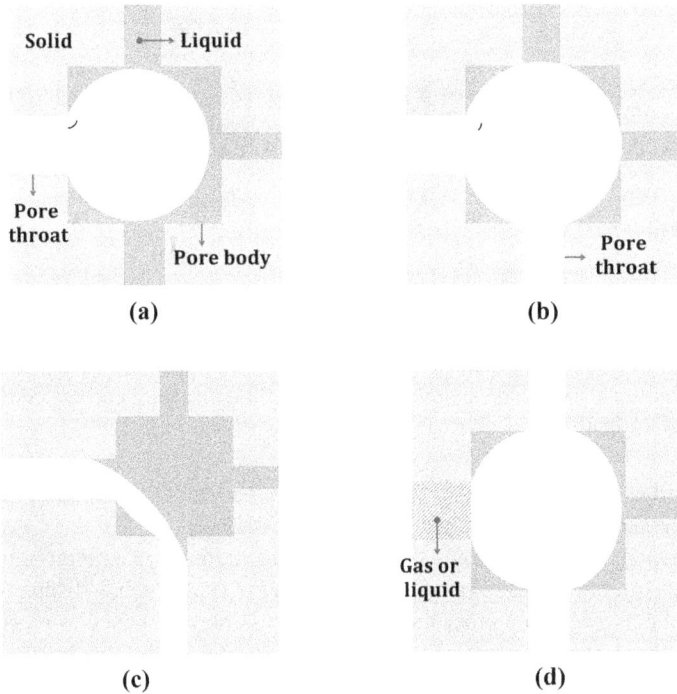

FIGURE 3.22 Schematic of the meniscus configuration in a partially filled pore body with (a) only one adjacent empty pore throat, (b and c) two adjacent empty pore throats neighboring to each other, and (d) two adjacent empty pore throats opposite to each other.

pore bodies in the pore network in our experiment, the capillary instability induced receding menisci will not invade the gas-filled pore body (because of the CVE), and the bubble shown in Figure 3.14 will not form.

To form the gas bubble shown in Figure 3.14, an empty pore body connected to at least two empty pore throats must be refilled completely by the liquid (an empty pore is the one occupied by gas), e.g., the central pore body in Figure 3.14 (i.e., pore body 3 in Figure 3.2). Refilling of a pore body with liquid is affected by the state (empty or filled) of the connected pore throats, as illustrated in Figure 3.22. A filled pore is the one filled with liquid. For an empty pore body connected to three filled pore throats and one empty pore throat (Figure 3.22a), the gas–liquid meniscus can remain concave toward the gas phase during refilling of a pore body, e.g., pore body 1 in Figure 3.18.

For an empty pore body connected to two filled pore throats neighboring to each other and two empty pore throats (Figure 3.22b and c), the meniscus also can remain concave toward the gas phase during refilling of the pore body; but the curvature radius can increase to a large value; see refilling of the central pore body in Figure 3.14 (i.e., pore body 3 in Figure 3.18). For an empty pore body connected to two empty pore throats opposite to each other (Figure 3.22d), the meniscus will change from concave to convex (toward the gas phase) in order to invade the pore

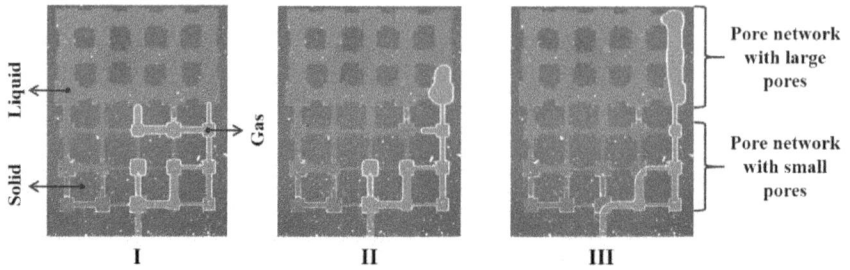

FIGURE 3.23 Capillary instability induced gas–liquid displacement in a microfluidic pore network with heterogeneous pore structures during evaporation. This composite pore network is composed of a zone with smaller pores and a zone with large pores. When the meniscus reaches to the zone where large pores prevail, the capillary instability occurs. Such capillary instability is attributed to the heterogeneity of the pore structure.

body, mainly attributed to the CVE induced by the sudden geometrical expansion between the pore body and the two empty pore throats.

Our experimental studies show that during slow drainage in a porous medium, the increase of the curvature radius of the advancing meniscus (which displaces the wetting phase) can lead some other menisci to be unstable and hence to recede toward the non-wetting phase. The displacement of the advancing and the receding menisci depends not only on the interplay between them (e.g., the flow pattern between them) but also on the pore structure, wettability, and state (e.g., the CVE, the wettability heterogeneity, the corner films, number of adjacent pore throats filled with the wetting phase, and the residual wetting phase in the pore body). The main driving force for the capillary instability induced two-phase flow is the difference in the capillary forces between the advancing and the receding menisci, which is induced by the apparent contact angle heterogeneity in the pore network in the present study. The difference in the capillary forces between the advancing and receding menisci can also be attributed to the structural heterogeneity in the pore network; see the capillary instability induced gas–liquid displacement during evaporation in a pore network composed of a large pore zone and a small pore zone in Figure 3.23.

To form the gas bubble in the pore network shown in Figure 3.14, a gas (non-wetting phase) occupied pore body connected to as least two empty pore throats must be refilled completely by the liquid (wetting phase). This refilling of the pore body depends on the competition between the driving force for the displacement of the receding menisci and the invasion threshold pressure for the liquid to refill the pore body. Both the driving force and the pore threshold pressure depend on the wettability of the pore.

In the present study, we repeat the drying experiments four times. The first two experiments are performed with camera; in these two experiments, gas bubble formation is always observed. Then, we repeat the experiment twice by using the microscope to get images of high resolution so as to get the detailed evolution of the moving menisci in the pore network; however, the gas bubble formation is not observed, even though the capillary instability induced refilling of pores always occurs, Figure 3.18. The reason could be that after the first two experiments, the pore network is contaminated and the wettability of the pore is altered. Refilling of the pores observed in the experiments with the camera and the microscope is almost the same (comparing Figures 3.14 and 3.18), indicating that the change of the wettability of the pore network after the first two experiments could be small. But, if the driving force for the movement of the receding menisci and the threshold pressure to refill the central pore body in the pore network (i.e., pore body 3 in Figure 3.18) are comparable, then even a small variation in the pore wettability can change their relative order and hence affects the formation of the gas bubble. When receding meniscus B refills pore body 3 in Figure 3.18, the advancing meniscus A invades pore body 9. As shown in Figure 3.18, the pore throats adjacent to pore body 2 have similar contact angles; whereas pore throats 8 and 10 adjacent to pore body 9 have different contact angles, indicating that the wettability of pore body 9 is not uniform. To this end, even a small change of the wettability distribution could affect significantly the capillary pressure of the meniscus in pore body 9 and hence the driving force. From this point of view, it is necessary to consider the variation of the surface wettability so as to understand in detail the dynamics of the capillary instability induced two-phase displacement in porous media.

For the quasi-2D pore network used in the experiment, the coordination number of each pore body is 4, which may be less than that in 3D porous media. More than two menisci can be interplayed with each other in a 3D porous medium, much more complex than the interaction of two menisci in the quasi-2D pore network revealed in the present study. In spite of this, we conjecture that the underlying mechanisms for menisci movement in the 2D and 3D pore networks should be similar.

3.5 SUMMARY

The kinetics of evaporation in the microfluidic pore networks are investigated. To disclose the role of the corner liquid films, two types of microfluidic pore networks are employed. For evaporation in the PDMS-based pore network, the corner liquid films cannot be formed. For drying in the silicon-glass based pore network, the corner liquid films can be found. Based on the visualization experiments with the microfluidic pore networks, the movement of the gas–liquid interface is clearly captured, and the CVE, the capillary scissors effect, and the capillary instability effect are revealed. The CVE is induced by the sudden geometrical expansion at the interface between a small pore and a large pore. Such geometrical expansion can hinder the movement of the triple line and hence seems to act as a valve. The capillary scissors effect is due to the local convex topology of the solid matrix. Such convex topology of the solid matrix can cut off the corner liquid films and hence seems to act as scissors. The capillary instability effect is induced by the heterogeneity in wettability and structure of the porous media. Naturally, the pore geometry considered in our microfluidic network, characterized by a sudden expansion at the throat pore–pore body junction network exacerbates some of the effects illustrated in this chapter. In real porous media, the local geometry can be quite different. Nevertheless, as illustrated throughout this chapter, microfluidic pore networks are quite valuable devices to study elementary phenomena occurring at the pore scale during drying.

REFERENCES

Armstrong, R.T., Berg. S. 2013. Interfacial velocities and capillary pressure gradients during Haines jump. *Physical Review E* 88: 043010.

Atefi, E., Mann, J.A., Tavana, Jr.H. 2013. A robust polynomial fitting approach for contact angle measurements. *Langmuir* 29: 5677–5688.

Canny, J.F. 1986. A computation approach to edge detection. *IEEE Transaction on Pattern Analysis and Machine Intelligence* 8: 679–698.

Chernov, N., Lesort, C. 2005. Least squares fitting of circles. *Journal of Mathematical Imaging and Vision* 23: 239–252.

Cieplak, M. Robbins, M. Influence of contact angle on quasistatic fluid invasion of porous media. *Physical Review B* 41: 11508.

Edery, Y., Berg, S., Weitz, D. 2018. Surfactant variations in porous media localize capillary instabilities during Haines jump. *Physical Review Letters* 120: 028005.

Gander, W., Golub, G.H., Strebel, R. 1984. Least squares fitting of circles and ellipses. *BIT Numerical Mathematics* 34: 558–578.

Mahmud, W.M., Naguyen, V.H. 2006. Effects of snap-off in imbition in porous media with different spatial correlations. *Transport in Porous Media* 64: 279–300.

Mayer, R.P, Stone, R.A. 1965. Mercury porosimetry – breakthrough pressure for penetration between packed spheres. *Journal of Colloid Science* 20: 893–911.

Odier, C., Levache, B., Santanach-Carreras, E., Bartolo, D. 2017. Forced imbibition in porous media: a fourfold scenario. *Physical Review Letters* 119: 208005.

Princen, H.H. 1969. Capillary phenomena in assemblies of parallel cylinders. *Journal of Colloidal and Interface Science* 30: 69–75.

Tao, W., Zhong, H., Chen, X., Selami, Y., Zhao, H. 2018. A new fitting method for measurement of the curvature radius of a short arc with high precision. *Measurement Science and Technology* 29: 075014.

Wong, H., Morris, S., Radke, C.J. 1992. Three-dimensional menisci in polygonal capillaries. *Journal of Colloid and Interface Science* 148: 317–336.

Wu, R., Kharaghani, A., Tsotsas, E. 2016. Capillary valve effect during slow drying of porous media. *International Journal of Heat and Mass Transfer* 94: 81–86.

Wu, R., Zhao, C.Y., Tsotsas, E., Kharaghani, A. 2017. Convective drying in thin hydrophobic porous media. *International Journal of Heat and Mass Transfer* 112: 630–642.

Wu, R., Zhang, T., Ye, C., Zhao, C.Y., Tsotsas, E., Kharaghani, A. 2020. Pore network model of evaporation in porous media with continuous and discontinuous corner films. *Physical Review Fluids* 5: 014307.

4 A Mesoscopic Approach for Evaporation in Capillary Porous Media
Shan Chen Lattice Boltzmann Method

Debashis Panda and Shubhani Paliwal
Birla Institute of Technology and Science,
Pilani-Hyderabad Campus
Imperial College London

Supriya Bhaskaran
Birla Institute of Technology and Science,
Pilani-Hyderabad Campus
Otto-von-Guericke-Universität Magdeburg

Githin Tom Zachariah
Bernal Institute, University of Limerick

Evangelos Tsotsas and Abdolreza Kharaghani
Otto-von-Guericke-Universität Magdeburg

Vikranth Kumar Surasani
Birla Institute of Technology and Science,
Pilani-Hyderabad Campus

CONTENTS

DOI: 10.1201/9781003011811-4

4.1   INTRODUCTION

4.1.1   Evaporation in Porous Media

Evaporative drying of porous media is an important unit operation in solids and semi-solids processing industries. Its application benefits preservation of food and improves directional stability of wood and handling of bulk drugs and food (Tsotsas et al., 2010). It involves displacement of wetting phase (usually water) by non-wetting fluid (air) under the simultaneous thermodynamic phase change of moisture evaporation at the interface. The displacement involves coupled heat and mass transfer in the complex void space, which makes evaporative drying in porous medium an intricate phenomenon. Moreover, the complexity of phenomenon comes from the interaction of forces such as capillary, viscous and gravity at the pore scale. The evaporation from wet porous material is economically necessary, but it is still a challenging scientific field to study even after the development of advanced experimental techniques and theoretical models. Since the experimental facilities are limited and expensive, several theoretical models have been developed to understand the complex multiphase phenomena in order to design efficient processes and products of higher quality.

Classically, mathematical models for evaporation in porous media are based on fictitious continuum for which conservation equations are derived by either homogenization or volume averaging (Quintard and Whitaker, 1993; Sahimi, 2011). The lack of length scale separation and the need for effective transport parameters are major limitations of continuum models. Pore network models (PNMs) at pore scale can describe the transport phenomena of evaporation in porous media and does not require any effective parameters (Laurindo and Prat, 1996; Lenormand et al., 1988; Metzger et al., 2007; Surasani et al., 2008, 2008a). The first basic PNM was proposed by Prat (1992) to predict the phase distributions and the corresponding evaporation rates. In his work, the evaporative drying algorithm is followed by dynamic invasion percolation rules. The PNM for evaporative process in porous media has expanded from a simple dynamic phase distribution to inclusion of gravity (Laurindo and Prat, 1996; Prat and Bouleux, 1999), viscous (Metzger et al., 2007; Yiotis and Boudouvis, 2004), thermal (Surasani et al., 2010, 2008b) and film effects (Chauvet et al., 2009; Prat, 2007) to obtain a robust algorithm. Along with the effects of thermal, viscous, capillary and gravity, the fluid–structure interactions also play a major role during evaporation in porous media. The damaging effects of capillary forces in particle aggregates are investigated by coupling of PNM with discrete element method (DEM) by Kharaghani et al. (2012). A PNM that accounts for the capillary valve effect (CVE) during evaporation in porous media has been implemented by Wu et al. (2016). The recent reconciliation of PNM with continuum approaches makes the model a success after relentless contribution from various scientific groups for the last three decades. However, PNMs are still an approximation of a real solid matrix where either the volume of pores is not considered or if considered, the pore and throats lack the irregularities of pore structures which play a major role in edge effects, Haines jumps, etc.

4.1.2 Lattice Boltzmann Method for Transport Phenomena in Porous Media

In the last few decades, lattice Boltzmann methods (LBMs) (Al–Ghoul et al., 2004; Chopard and Hoekstra, 2004) emerged as a direct numerical simulator for the multiphase flows in the porous medium. The LBM has major advantages over traditional pore-scale modelling techniques. Firstly, it emerges from the Boltzmann kinetic molecular dynamics – a more foundational entity than a conventional scaled modelling technique. In other way, one can view the LBM as a mesoscopic approach by incorporating the interparticle forces. This perception holds LBM a basic yet a better method for implementing surface tension, capillary regime, thermal effects and fluid–solid interactions. Secondly, the pressure in the LBM is expressed directly from the non-ideal equation of states for the multiphase fluids. Hence, the pressure is obtained as a function of density in the LBM unlike other techniques, where the Poisson equation (computationally expensive) is required to be solved to obtain the pressure field. Thirdly, the no-slip boundary condition is easily implemented in the solid domain. Thus, it becomes easier to handle complex geometry like void space in porous medium. Finally, most of the operation in the LBM is local to a given spatial node. Therefore, the LBM is easily parallelizable too.

LBM rapid development in the last two decades is the reason of its popularity in the multiphase flows in the porous medium. The multiphase flow in porous media using LBM was introduced by Martys and Chen (1996). Chin et al. (2002) used

LBM for various viscosity ratios to model viscous fingering, Kang et al. [37] investigated the capillary number (Ca) and wettability for viscous fingering. A representative elementary volume-scaled LBM is implemented by Abrach et al. (2013), and El Abrach et al. (2013) for the evaporation in porous media. A multiphase LBM for incorporating the pore-scale physics during the evaporation process in porous medium was first introduced by Zachariah et al. (2019). They investigated Haines jumps and threshold pressure build-up qualitatively during evaporation in capillary porous media. Panda et al. (2020b) recently extended the works to a quantitative approach by understanding the micro–macro interactions in a bundle of capillaries. In this chapter, we introduce the pore-scale physics during evaporation in porous media using a popular direct numerical simulator, i.e., LBM. LBM is a multiphase flow simulator for various advantages, i.e., (a) automatic interface tracking method, (b) easily parallelizable, (c) no requirement of solving the Poisson Equation and (d) interlinks between macroscopic continuum properties and microscale intermolecular dynamics.

4.1.3 OUTLINE

Section 4.2 presents the description of the SC LBM for the evaporation in porous media. Section 4.3 discusses the pore-scale physics of during evaporation phenomenon. The dominant forces namely competitive capillary and viscous forces are presented in Section 4.3.1. The pore-scale physics such as CVE, pore invasion patterns and Haines jump is presented in Section 4.3.2. In Section 4.3.3, the macroscopic evaporation kinetics and the micro–macro interactions of the complex multiphase fluid dynamics by traversing from one-dimensional bundles to complex porous structures are presented. We examine the film effects in the phenomenon in Section 4.3.4. Finally, in Section 4.4, we present a blueprint of the outlook in the advancements of LBM for evaporation in porous medium.

4.2 LATTICE BOLTZMANN METHOD: EVAPORATION IN POROUS MEDIA

4.2.1 GENERAL LATTICE BOLTZMANN METHOD

Particle Distribution Function (PDF) f is the basic constituent of LBM. It quantifies the number of particles with a particular velocity in a defined position at any time. Figure 4.1a represents a typical computation domain with random porous medium during evaporation using LBM simulation. Grey, black and white colours in the computation domain represent the solid, the liquid and the gas nodes respectively. Convective flow conditions lead to the formation of this boundary layer, see dotted lines in Figure 4.1a. The 2D arrangement of lattice nodes in the gas phase and D_2Q_9 representation of discrete velocity sets in every lattice are presented in Figure 4.1b and c respectively. The similar arrangements can be assumed in the bulk of the liquid and solid phase, and for the lattice nodes representation in multiphase, refer Figure 4.2.

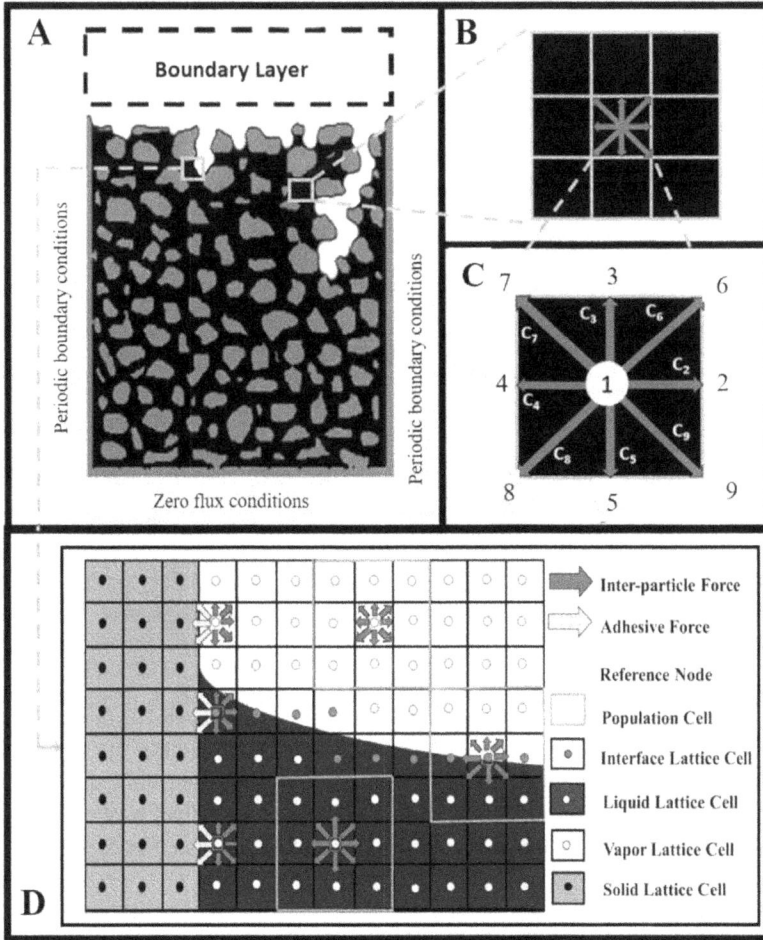

FIGURE 4.1 (a) Representation of 2D random porous medium undergoing LBM simulation. (b) 2D lattice nodes representation. (c) D_2Q_9 representation of lattice node and (d) schematic representation of solid–liquid–vapour interaction in Shan Chen LBM. (Adapted from Zachariah et al. (2019).)

Table 4.1 presents the mathematical representation of the Shan Chen LBM (SC LBM) for evaporation in capillary porous medium. The evolution of PDF at each lattice node is represented using the Boltzmann equation (Eq. 4.1) (Tolman, 1922), where f_k is the PDF in the kth direction, $\vec{c}_k$ is the discrete velocity in the kth direction and $\vec{F}_k$ is the forcing term. The first term Ω_k on the RHS of Eq. (4.1) is the collision operator, where τ is the relaxation time and f_{eq} is the equilibrium PDF. The collision scheme used in Eq. (4.2) is popularly known as Bhatnagar–Gross–Krook (BGK) (Zou et al., 1995) representation. Alternative collision operators can be implemented for less spurious currents and stability, e.g., Multiple Relaxation Time (MRT) (McCracken and Abraham, 2005; Monaco et al., 2011; Yoshida and Nagaoka,

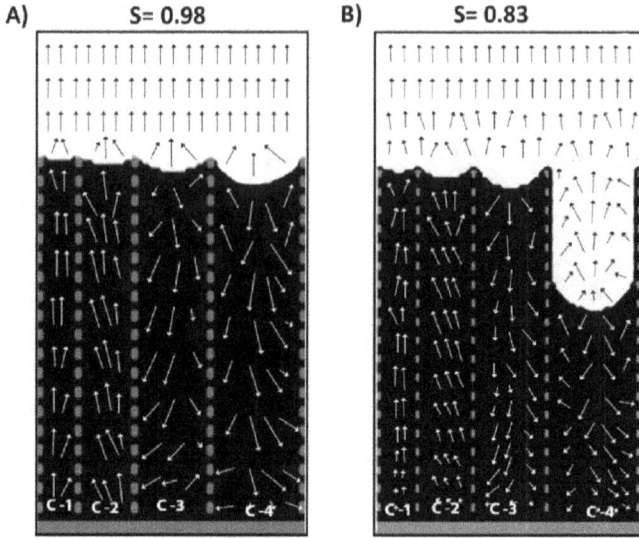

FIGURE 4.2 LBM simulation for isothermal evaporation in bundle of capillaries at $S = 0.98$ and $S = 0.83$ with quiver plots of the coupled liquid and vapour transport. (Adapted from Panda et al. (2020b).)

2010) collision operators, which is more accurate. However, MRT is computationally demanding than the BGK collision operator. Thus, for maintaining the computational efficiency, in a diffusion-dominated regime as in isothermal condition, where relaxing towards equilibrium (as in BGK collision) is a physical way of understanding the diffusion process, the BGK collision operator can be accepted in permissible limits.

The equilibrium PDF f_{eq} derived from Maxwell–Boltzmann distribution function is calculated as shown in Eq. (4.3). Various discretization velocity sets have been implemented for two-dimensional domains (D_2Q_4, D_2Q_5, D_2Q_7 and D_2Q_9) among which D_2Q_9 (two-dimensional nine velocities) is used in this study. Figure 4.1c presents the applied D_2Q_9 (Perumal and Dass, 2015) in the LBM simulations, where the weights w_k and the discrete directional velocities $\vec{c}_k$ can be written as shown in Eqs. (4.4) and (4.5). The Boltzmann Equation is discretized with respect to space, velocity and time using the D_2Q_9 velocity set to obtain Lattice Boltzmann Equation as shown by Eq. (4.5). The density and velocities for each node are calculated using Eqs. (4.6) and (4.7).

4.2.2 SHAN CHEN LATTICE BOLTZMANN METHOD

In the evaporation process, the liquid wets the porous media as it is held due to capillarity, while the gas phase invades inside the porous media as time progresses. In a diffusion-dominated convective evaporation process, the formed vapour diffuses through the gas phase towards a moisture-less boundary. Thus, it is understood that counter-diffusion of vapour and gas occurs during evaporation process, while the

TABLE 4.1
The LBM Equations for Evaporation in Porous Media

Equation		Expression	Eq. (#)
Mesoscopic Transport Equations	Boltzmann equation	$\dfrac{\partial f_k}{\partial t} + \vec{c}_k\,\dfrac{\partial f_k}{\partial x} = \Omega_k\left(f\right) + \vec{F}_k$	Equation (4.1)
	Collision operator	$\Omega_k\left(f\right) = -\dfrac{1}{\tau}\left(f_k - f_k^{eq}\right)$	Equation (4.2)
	Equilibrium particle distribution function	$f_k^{eq} = w_k\rho\left(1 + \dfrac{\vec{c}_k.\vec{u}}{c_s^2} + \dfrac{(\vec{c}_k.\vec{u})^2}{2c_s^4} - \dfrac{\vec{u}.\vec{u}}{2c_s^2}\right)$	Equation (4.3)
Lattice node Discretization using D2Q9	Weights	$w_k = \begin{cases} 4/9 & k=1 \\ 1/9 & k=2,3,4,5 \\ 1/36 & k=6,7,8,9 \end{cases}$	Equation (4.4)
	Discrete velocity set	$\vec{c}_k = \begin{cases} (0,0) & k=1 \\ (\pm1,0),(0,\pm1) & k=2,3,4,5 \\ (\pm1,\pm1) & k=6,7,8,9 \end{cases}$	Equation (4.5)
	Macroscopic density	$\rho = \displaystyle\sum_{k=1}^{9} f_k$	Equation (4.6)
	Macroscopic momentum	$\rho\vec{u} = \displaystyle\sum_{k=1}^{9} f_k\vec{c}_k$	Equation (4.7)
	Lattice Boltzmann equation	$f_k\left(\vec{x}+\vec{c}_k\Delta t, t+\Delta t\right)$ $= f_k\left(\vec{x},t\right) - \dfrac{\Delta t}{\tau}\left(f_k\left(\vec{x},t\right) - f_k^{eq}\left(\vec{x},t\right)\right) + \vec{F}_k$	Equation (4.8)
Shan-Chen Lattice Boltzmann Equations	Interparticle forces	$\vec{F}_{int}\left(\vec{x},t\right) = -G\psi\left(\vec{x}\right)\displaystyle\sum_{k=1}^{9} w_k\psi_k\left(\vec{x}+\vec{c}_k\Delta t\right)\vec{c}_k$	Equation (4.9)
	Solid–fluid interaction	$\vec{F}_{ads}\left(\vec{x},t\right) = -G_{ads}\psi\left(\vec{x}\right)\displaystyle\sum_{k=1}^{9} w_k\psi_k^{wall}\left(\vec{x}+\vec{c}_k\Delta t\right)\vec{c}_k$	Equation (4.10)
	Water–air interactions	$\vec{F}_a\left(\vec{x},t\right) = -G_{ab}\psi^a\left(\vec{x}\right)\displaystyle\sum_{k=1}^{9} w_k\psi_k^b\left(\vec{x}+\vec{c}_k\Delta t\right)\vec{c}_k$	Equation (4.11)
	β-scheme	$\vec{F}_{int}\left(\vec{x},t\right) = c_0\beta G\vec{\nabla}\,\psi\left(\vec{x}\right) - \dfrac{1-\beta}{2}\,c_0 G\vec{\nabla}\,\psi^2\left(\vec{x}\right)$	Equation (4.12)
	Pseudopotential calculation	$\psi\left(\rho\right) = \sqrt{\dfrac{2\left(P-\rho c_s^2\right)}{c_0 G}}$	Equation (4.13)
	Carnahan Starling equation of state	$P = \rho RT\,\dfrac{\left(1+\dfrac{b\rho}{4}+(\dfrac{b\rho}{4})^2-(\dfrac{b\rho}{4})^3\right)}{(1-\dfrac{b\rho}{4})^3} - a\rho^2$	Equation (4.14)
	Velocity shifting	$\vec{u}^{eq} = \vec{u} + \dfrac{\vec{F}_{int}}{\rho}$	Equation (4.15)
	Multicomponent velocity	$\vec{u}' = \dfrac{\displaystyle\sum_\sigma \dfrac{1}{\tau_\sigma}\displaystyle\sum_k f_k^\sigma \vec{c}_k}{\displaystyle\sum_\sigma \dfrac{\rho_\sigma}{\tau_\sigma}}$	Equation (4.16)
	Diffusion model	$f_k^{eq} = w_k\rho$	Equation (4.17)
	Vapour transport equation	$\dot{M}_{ij} = -\dfrac{A_{ij}\delta\tilde{M}_v P_g}{L\,RT_{avg}}\ln\left(\dfrac{P_g-P_{v,i}}{P_g-P_{v,j}}\right)$	Equation (4.18)

meniscus recedes by following the interfacial rules depending on the structural and thermal properties. Therefore, it is considered as a complex multiphase problem of air–liquid–vapour–solid interface. Thus, to account the complex interfacial effects in the phenomena, two PDFs are introduced for air and water. While the PDF corresponding to air diffuses inside the porous medium, the PDF corresponding to water undergoes a phase change and the vapour formed diffuses towards the top of the boundary layer. Various multiphase LBMs have been developed, e.g., free energy LBM (Zacharoudiou and Boek, 2016), Rothman–Keller LBM (Huang et al., 2013) and the very popular SC LBM (Chibbaro et al., 2008). Even though the Rothman–Keller is a better approach for air–water simulations but the SC LBM is chosen as both the air–water as well as the phase change behaviour can be utilized in permissible limits. Moreover, the simplicity to understand the inter-particle interactions in SC LBM enables one to understand the micro–macro interactions in the evaporation process in capillary porous media.

4.2.2.1 Liquid–Vapour Interaction

In SC LBM (Liu et al., 2016; Shan and Chen, 1994), the introduction of pseudo-potential to incorporate forcing terms in the model perturbs the ideal equation of state. Thus, it helps to attain phase change behaviour in the model. Some researchers (Colosqui et al., 2012; Yuan and Schaefer, 2006) easily implement the forcing terms to attain higher density ratio stability by a simple strategy of shifting the velocity in f_{eq} to the incorporated momentum due to inter-particle force as shown in Eq. (4.8). Further, the inter-particle forces are provided as follows: for a multiphase system, the intermolecular forces are additive in a real vector space and occurs in pair of two molecules. Henceforth, a higher density space is clearly understood to have higher intermolecular forces. Thus, the forces are regarded to be proportional to the product of the density of a node and its neighbouring node. Next, we introduce a kernel function G which is a signum function to rule the distance between the considered node and its neighbouring node. Lastly, the total inter-particle force is an integral of all the neighbouring products of G and density of the node and the neighbouring node. However, the assumption of two-particle interactions put a constraint over realistic imitation of intermolecular forces. Therefore, an effective density called pseudopotential (ψ) replaces the density in the above argument. For the numerical artifact, the total force is demonstrated as a sum of all local force vectors of a corresponding node given by Eq. (4.9). The attraction–repulsion strength of the inter-particle forces is tuned by the constant parameter G which is the result of the signum function discussed earlier. The non-ideal equation of state for the phase change is given in Eq. (4.10). In the D_2Q_9 velocity set, c_0, c_s is taken as 6 and $1/\sqrt{3}$ respectively (Lallemand and Luo, 2000). ψ is calculated using the equation of state (EOS) in Eq. (4.11) which is the other way around to write Eq. (4.10).

Figure 4.1d represents the types of inter-particular forces where three different populations are represented, i.e., liquid–liquid, vapour–vapour and interface nodes (Zachariah et al., 2019). From the above discussion, we can confirm that the total inter-particle force at a node in pure liquid or gas or vapour is 0 and the imbalance of inter-particle force at the interface leads to the formation of a diffusive interface between the liquid and the gas. Therefore, SC LBM automatically tracks the interface, unlike other interface tracking methods. In our work, we have implemented Carnahan Starling equation of state (CS EOS), given in Eq. (4.11), where a and b are taken as 1 and 4 respectively. The CS EOS is applied because it generates very low spurious currents and is applied for a wider temperature and density ranges (Yuan and Schaefer, 2006). Moreover, the force discretization plays a vital role in the attainment of high-density ratio multiphase LBM. Therefore, combining Eq. (4.10) with local interparticle forces and permissible numerical approximation of the gradient, we implement Eq. (4.13), where β is a tuning factor. For $\beta = 1$, Eq. (4.14) coincides with Eq. (4.10), and for $B = 0.5$, it coincides with mean-value approximation (Kupershtokh et al., 2009).

4.2.2.2   Solid–Fluid Interactions

Solid–fluid interaction can be incorporated by modifying Eq. (4.9) to calculate the fluid–solid adhesion given by Eq. (4.10), where G_{ads} is the solid–fluid interaction strength constant and ψ_{wall}^{k} is the solid nodes presence indicator which is 1 for solid node and 0 for fluid. Figure 4.1d represents a set of nodes with liquid–gas interface adjacent to a wall node. The interface lattice node adjacent to the solid wall is surrounded by four types of lattice nodes, i.e., at least one interface lattice node and the remaining from solid, liquid and vapour lattice nodes. This creates variation in effective mass of the interface nodes which leads to a competition between adhesive (green) and cohesive (yellow) forces. This competitivity result in the meniscus formation and water hydrophilicity is developed.

4.2.2.3   Water–Air Interactions

Inter-particular force in multicomponent Shan Chen model is given by Eq. (4.15), where a and b denote the two components. However, in our model with a constant temperature, the force strength between the two components (i.e., water and air) is assumed to be zero (G in Eq. (4.16) is zero). This assumption results in an interesting scenario in which evaporation via vapour transport is done by momentum transfer. Moreover, the computational cost for simulations also decreases. However, modified multicomponent macroscopic velocity is applied to keep the multicomponent effect in the system.

4.2.3   Initial and Boundary Conditions

The densities of the liquid, vapour and air (ρ_l, ρ_v, ρ_a) are 990, 0.93 and 1.5 kg/m³ respectively. These are converted from the lattice to metric units by using the reduced properties of water and air. The collision relaxation time parameters for both air and water (τ_g and τ_w) are taken as unity. The boundary condition applied to the model is similar to our previous work (Zachariah et al., 2019). Periodic boundary conditions are applied at the lateral sides of the boundary layer and Neumann open boundary condition is kept for the top layer of the domain. The convective boundary condition is more stable and accurate, but the implementation of Neumann boundary condition is easier (Lou et al., 2013). The derivative of variables is set to zero in this boundary condition. Further, zero moisture condition is implemented in the upper boundary by setting the out-streamed particle distribution function as zero. Bounce-back boundary condition is implemented to keep no-slip boundary condition at the solid nodes (Zou and He, 1997).

4.2.4   IMPLEMENTATION OF ISOTHERMAL EVAPORATIVE IN SC LBM

The proposed model is implemented with isothermal conditions as the temperature in the EOS (Eq. 4.12) is kept constant. The zero-moisture boundary condition is implemented at the top of the boundary layer to explicitly keep a diffusion model in the boundary layer to achieve evaporative drying. To implement the explicit diffusion model, the equilibrium PDF (f_{eq}) for the boundary layer is kept as (Mohamad, 2011) given in Eq. (4.17).

4.2.5   LATTICE BOLTZMANN METHODOLOGY

The local forcing calculations for the interfacial effects are the major concern in any multiphase flow model. In LBM, for an accurate imitation of permissible–diffusive interface, we implement liquid–vapour, liquid–liquid, vapour–vapour, liquid–air, vapour–air and air–air cohesive forces and liquid–solid, vapour–solid and air–solid adhesive forces. These cohesive and adhesive forces lead to a stabilizing three-phase solid–liquid–gas contact line. For more understanding of the local forces leading to interfacial effects, one can refer Figure 4.1. The Lattice Boltzmann Equation in Eq. (4.8) is implemented as a two-step algorithm as follows:

Collision:

$$f_k^*(\vec{x},t) = f_k(\vec{x},t) - \frac{\Delta t}{\tau}\left(f_k(\vec{x},t) - f_k^{eq}(\vec{x},t)\right) + \vec{F}_{k,int} \qquad (4.19)$$

Streaming:

$$f_k(\vec{x}+\vec{c}_k\Delta t, t+\Delta t) = f_k^*(\vec{x},t) \qquad (4.20)$$

where $f_k(\vec{x}, t+\Delta t)$ is the post collided PDF. Then, the post-collided PDFs are streamed to the neighbouring nodes to complete the streaming step. A pair of these two steps are considered to be one time-step for the LB simulation. In the above expression, the interparticle forces are calculated using Eq. (4.12), where G is taken as −1 to keep the expression positive under the root in Eq. (4.13). However, stabilized discretization needs higher order numerical approximation of the gradient which is accompanied by taking $\beta = 1.16$ for CS EOS. The incorporated force is introduced in f_{eq} by velocity shifting method as shown in Eq. (4.15). Further details of the algorithm with accuracy check are shown in our previous work (Panda et al., 2020b).

4.3　RESULTS AND DISCUSSION

The evaporation in porous medium at pore scale is a demanding phenomenon and upscaling the micro-dynamics to the macroscopic evaporation kinetics is still a challenging field of interest. The coupling of liquid transport and vapour transport during the dynamic evolution of spatio-temporal invasion patterns is needed to be elucidated in detail. In this section, we present the fundamental aspects of evaporation at pore scale. The structure of the elucidation is as follows: firstly, we present the micro–macroscale interactions in the porous media during the evaporation in porous media (Section 4.3.1). The micro–macro interactions are first studied with the help of the single-dimensional bundle of capillaries (Section 4.3.1.1). Secondly, we extend the concepts to regular and irregular porous media (Section 4.3.1.2). Further, we describe the physics of the pore-scale events which is primarily dominated by the structure of the porous media (Section 4.3.2). Finally, we present one of the major outcomes of the work, i.e., the film flows during the evaporation in porous media (Section 4.3.3).

4.3.1　DOMINANT PHENOMENON

A typical strategy is employed to understand the evaporation phenomenon, i.e., by relaxing the constraints of invasion patterns step-by-step from simple to complex random solid matrix geometry. To start with a simple geometry, a bundle of capillaries is an excellent choice because (i) it is considered as a one-dimensional representation of the porous media, (ii) the horizontal transport effects are reduced to an extent that its sole purpose is to keep inter-channel connectivity – thus, the liquid transport is dependent upon vertical transport effects. (iii) Such imitation of the porous media orchestrates to a desired model to understand the micro–macro interactions in the evaporation phenomena.

　　For the purpose, a 4-bundle of capillaries **C1, C2, C3** and **C4** are introduced in increasing order of the size (i.e. 5, 15, 20 and 25 μm). The boundary layer is taken to be 100 μm. The simulation is evaluated at atmospheric pressure and isothermal temperature of 60°C. The contact angle is set at 37° to show that the capillary and viscous forces are competitive in nature. At $S=0.98$ and $S=0.83$, the evaporation in bundle of capillaries is shown in Figure 4.2. The invasion patterns are plotted with the black region signifying the liquid phase, the white region signifying the gas phase and the grey region signifying the solid phase. Both liquid transport and vapour transport coupling are plotted as quivers with white and black colour respectively. The diffusion dominant boundary layer induces a vertical vapour transport linking between the evaporating meniscus and the top of the boundary layer.

　　The liquid transport conveys the fact that the water is transported from the bundle C-4 and C-3 to C-2 and C-1. It is for the following reasons: The size heterogeneity induces difference in capillary pressure at each void space. The capillary pressure is given by $P_c = P_g - P_l = \dfrac{2\sigma}{r}$, where P_c is the capillary pressure, P_g is the gaseous pressure and P_l is the liquid pressure. The surface tension σ is constant as isothermal condition is assumed. Thus, the capillary pressure P_c depends upon width of the void space, i.e., wider void space shows lower capillary pressure. Moreover, capillary pressure also equals to the difference of the non-wetting and wetting fluid

$(P_g - P_l)$. The non-wetting pressure is constant and equals to atmospheric pressure. Therefore, lower capillary pressure results in a higher liquid pressure in the interface. It induces a gradient within the capillary pressure where the liquid is transported from the wider void space to narrower void space. At capillary regime, it is called capillary pumping phenomenon where the capillary heterogeneity dominates the liquid transport in the porous media. Therefore, the micro-capillaries (i.e., **C1, C2, C3**) are pinned to the surface, while the capillary **C4** is receding. The pinning helps to vapour transport from the surface of the bundle of capillaries.

4.3.2   PORE-SCALE PHYSICS

In Section 4.3.1 we have discussed the pore-scale dynamics of liquid and vapour transport using bundle of the capillary porous media. However, such imitation comes with various shortcomings. Firstly, it reduces a porous media representation to a vertical transport length model and discards the horizontal liquid transport in the porous media. Secondly, the solid representation is reduced to the minimal for defining the capillary pumping in the porous media. The minimization of the solid representation hides various important pore-scale dynamics that are ubiquitous during the evaporation in porous media. In this section, we present a two-dimensional representation of the porous media to introduce various pore-scale dynamics during isothermal evaporation in porous media. For the purpose, we present a pore network with circular pore space. The average throat radius of the pore network is 40 μm and a standard deviation of 6 μm. The thermodynamic conditions are similar to the bundle of capillaries discussed in Section 4.3.1

4.3.2.1   Capillary Valve Effect

The use of bundle of capillaries deduces that "in an isothermal condition, the macro-throats open towards the gas phase invade first". Therefore, as shown in Figure 4.3(i), the macro-throats recede, while the micro-throats wet the surface of the pore network. However, a peculiar observation can be scrutinized at the pore opening. The meniscus takes a halt at the pore-space opening. As shown in Figure 4.3(i), the throats signified in the white box are receding, while the meniscus signified in the grey circle is still at halt. Following is the explanation of the halt at the pore opening: A threshold pressure is defined for the invasion of the throat. It is the minimum pressure to invade a throat, which is greater than the capillary pressure at equilibrium. The imbalance of cohesive and adhesive forces at the solid–liquid–gas contact line proves that the capillary pressure is also dependent on the contact angle (i.e., $\cos\theta$). As a meniscus recedes in a throat, the invasion follows a constant contact angle. However, when it reaches the pore-space opening, it experiences a change in the contact angle due to sudden enlargement of the pore space (Panda, 2020a). The change (sudden increase) in the contact angle explains an increase in the threshold pressure to invade. The building up of pressure to surpass the threshold pressure thus leads to a halt at the pore-space opening. This phenomenon is called "Capillary Valve Effect (CVE)". The CVE plays a major role in the invasion patterns during evaporation in porous media. For example, the halt leads to an opportunity for the horizontal throats to invade. If the horizontal throats won't perceive enough time to invade, then

I. Capillary Valve Effect

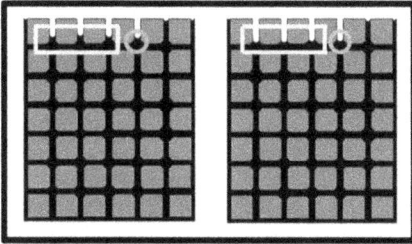

III. Threshold Pressure

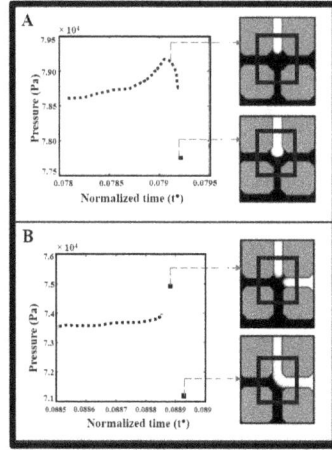

II. Burst and Merge Invasion

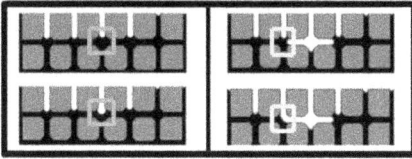

FIGURE 4.3 (i) Capillary calve effect (CVE) in pore space, (ii) burst and merge invasion and (iii) quantification of CVE by threshold pressure during evaporation in porous medium. (Adapted from Panda et al. (2020a).)

it results in higher number of disconnected clusters in the porous medium. Thus, a larger number of disconnected clusters are then affecting the evaporation kinetics.

4.3.2.2 Burst and Merge Invasions

The post-CVE affects the invasion of pores into two distinctive pore invasion patterns, which leads to abrupt fluctuations in liquid transport in the porous medium. The post-CVE refers to the fast invasion of the pore with higher threshold pressure by experiencing a contact angle instability in the junction. Such pore invasion from one throat imitates a bursting phenomenon and thus is called the "Burst" invasion as shown in grey box in Figure 4.3ii. As discussed, the CVE provides sufficient time for the horizontal throats to recede. In this case, it eventually leads to two menisci junction at one pore as shown in white box in Figure 4.3ii. This leads to "merging" of the menisci at the pore opening and thus is called "merge" invasion. The merge invasion plays a major role of imitating the realistic evaporation phenomena in porous media by reducing the number of disconnected clusters in the evaporation process.

The quantification of the above discussion is successfully captured in the LBM simulations as shown in Figure 4.3iii. It is clearly shown in the invasions patterns in a box (Figure 4.3iii) that the radius of curvature of merge invasion is greater than the burst invasion. Therefore, the threshold pressure for the burst invasion (7.92×10^4 atm) is higher than the merge invasion (7.5×10^4 atm). Another significant observation is the time lapses of the two kinds of pore invasion. The time lapse in the burst invasion ($\Delta t^* = 0.07928 - 0.07891 = 0.00037$) is higher than the merge invasion ($\Delta t^* = 0.08891 - 0.0889 = 0.00001$). It signifies that the merge invasion is faster than the burst invasion. The rapid merge invasion is due to two reasons: (i) The menisci junction coalescence dominates the liquid movement due to capillary forces and (ii) a larger radius of curvature attains lower threshold pressure to attain interfacial equilibrium.

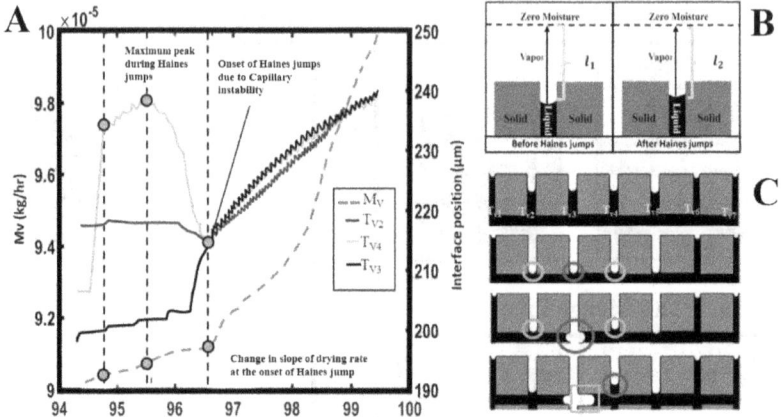

FIGURE 4.4   (a) Quantitative analysis of Haines jumps during isothermal evaporation of a plain-pore network, (b) schematic representation of the effect of Haines jumps to vapour transport during isothermal evaporation in porous medium and (c) qualitative analysis of Haines jumps during isothermal evaporation of plain pore network. (Adapted from Panda et al. (2020a).)

4.3.2.3   Haines Jumps

The sudden pore invasions lead to sudden pressure fluctuations in the liquid transport. These fluctuations are sometime so high that it leads to sudden refilling of the neighbouring throats. Such sudden refilling illustrates a jump at the interface, commonly called the "Haines jumps". In Figure 4.4c, it considers throats named as T_{v1}, T_{v2}, T_{v3}, T_{v4} and so on. As discussed, with time increments, the macro-throats start to invade until it observes CVE. In this process, the throat (for example, T_{v3} in Figure 4.4c) which attains the threshold pressure first gets invaded and, in this case, it considers a burst invasion. The burst invasion leads to a pressure fluctuation in the pore space. This sudden fluctuation induces opposite liquid pressure gradient for a back flow towards the neighbouring throats meniscus (for example, T_{v2} and T_{v4}). When the opposite liquid pressure gradient surpasses the invading pressure of the neighbouring throat, it observes a Haines jump. This phenomenon is described in Figure 4.4a. the three throats receded to 218 µm and the throat T_{v3} gets invaded first and recedes to 203.5 µm, while the throat T_{v2} jumps to 220 µm and the throat T_{v4} jumps suddenly to as high as 240 µm due to the combined effect of Haines jumps and invasion of neighbour horizontal throat to the pore.

The Haines jumps are a significant physical observation that directly affects the evaporation rates. This argument is clearly discussed using a pictorial representation in Figure 4.4b. It shows two snapshots distinctly to describe the before and after meniscus position during Haines jump event. In a diffusion-dominated evaporation, the vapour transport length is directly related to the vapour transport resistance, i.e., with decrease in the vapour transport length, vapour transport resistance also decreases. Let l_1 and l_2 be the vapour transport length before and after the event. Evidently, due to the Haines jump, $l_2 < l_1$ that means the vapour transport resistance

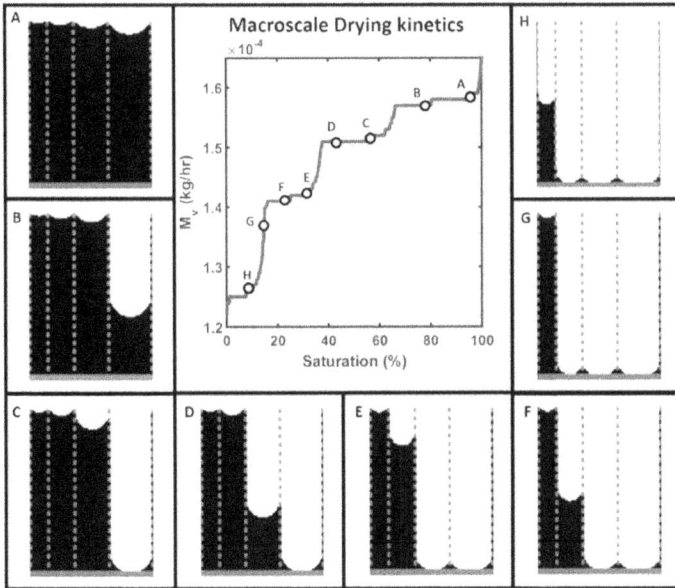

FIGURE 4.5 Micro–macro interactions in a bundle of capillaries. (Adapted from Panda et al. (2020b).)

decreases. It implies that the evaporation rate during this period increases. Therefore, it is observed in Figure 4.4a that when T_{v4} is experiencing Haines jump event, the evaporation rate slope increases in this time interval.

4.3.3 MICRO–MACRO INTERACTIONS

The major portion of the works are to understand the relationship between the micro-dynamics of the invasion to the upscaled evaporation kinetics. For the purpose, we develop a framework that one can exploit to understand the micro–macro interactions in the porous media. As discussed, we first introduce the concept of CRP and FRP using cyclic plots in 4-bundle of capillaries. Next, we orchestrate the ideas to an array of bundles to imitate a porous media. In the final stage, to reform the idea of micro–macro interaction, we present a real porous medium with random solid structure.

4.3.3.1 4-Bundle of Capillaries

As discussed in Section 4.3.1, we begin our discussion with the similar 4-bundle of capillary channels computational domain with the similar thermodynamic param-eters. Figure 4.5 represents the micro–macro interactions during evaporation in bun-dle of capillaries, i.e., M_v (evaporation rate) vs S (saturation). The system is initialized with full saturation $S = 100\%$ so that the initial contact angle of the water is 90°. The invasion patterns are shown in a circle from Figure 4.5a to Figure 4.5b. Hence, at the onset of the simulation, the first step is to stabilize the meniscus. The stabilization leads to an increase in the vapour transport length (L_T). Therefore, the rate of vapour transport decreases. The upscaling of the phenomena leads to an initial sudden fall in the evaporation kinetics (Figure 4.5a). The stabilization of the meniscus is then followed by the receding of the wider capillary channel **C4** as discussed in Section 4.3.1. While the receding leads to the capillary pumping of the neighbouring micro-throats, the other micro-throats are pinned to the top of the bundles. Hence, the vapour transport length in this period is held constant. Therefore, we observe a CRP from A to B (see Figures 4.5a and b). Once, the macro-throat is completely evaporated, the next macro-throat takes the charge of pumping the micro-throats. At this point, the vapour transport length observes a sudden increase due to falling of a new throat. Therefore, the evaporation rate observes a fall or FRP is observed (Figure 4.5c). Similarly, the CRP followed by an FRP is observed for **C3**. The cycle constant rate period (CRP) and falling rate period (FRP) is observed throughout the simulation as one macro-throat is enough to hold the micro-throats to the surface of the bundles. However, such cycle nature is not the accurate imitation of the evaporation in porous media. In order to understand the micro–macro interactions in the porous media, we extend our 4-bundle to an array of bundle of capillaries.

4.3.3.2 Array of Capillary Channels

In this section, we introduce the array of channels that imitates an evaporation kinetics of a porous media. In reality, the pore-size distribution is considered to be a bimodal distribution. However, the channel size distribution must be analysed for both mono- and bimodal channel size distribution. The micro-throats size is 6 ± 2 μm and the macro-throats size is 18 ± 6 μm. In the monomodal distribution, all throats are considered to be micro-throats; whereas in the bimodal distribution, a macro-throat is initiated at a regular interval of five throats. As discussed earlier, the kinetic and thermodynamic parameters are kept similar in the follow-ups with Section 4.3.1.

The normalized evaporation kinetics for both monomodal and bimodal distribution are shown in Figure 4.6. The invasion patterns for bimodal distribution at discrete saturation are shown in Figure 4.6a–f. At the onset of meniscus stabilization, the macro-throats start to recede and are preferentially empty. The preferential receding leads to the capillary pumping to the micro-throats in the bundles. Thus, we observe a constant vapour transport length during this period. The constant vapour transport length leads to constant evaporation rate in this period. This is called the CRP (Figure 4.6a–c). Once all the macro-throats are emptied, the absence of capillary pumping leads to the receding of the micro-throats (Figure 4.6d–f). This phenomenon is analogous to the disconnected clusters in the porous media. During this period, the vapour transport length increases with decreasing saturation of water. This is called the FRP. On the other hand, in a monomodal distribution, the absence of macro-throats leads to an absence of preferential emptying of the channels. Thus, the receding occurs only by the minor heterogeneity of the channel size which leads to an absence of CRP. At 60% of the evaporation kinetics, the macro-throats are sufficiently emptied. The further evaporation only depends upon the thermodynamic condition of the evaporation phenomena. Henceforth, we observe similar evaporation kinetics in both monomodal and bimodal after $S = 60\%$.

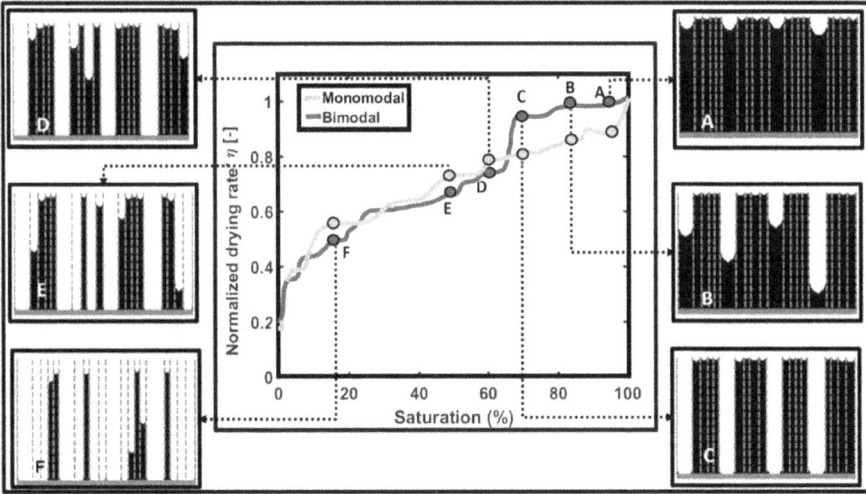

FIGURE 4.6 Micro–macro interactions in the array of bundle of capillaries. (Adapted from Panda et al. (2020b).)

4.3.3.3 Isothermal Evaporation in Random Porous Media

After establishing the basic concepts of micro–macro interactions in isothermal evaporation in bundles of capillaries, which are strictly a one-dimensional representation of a porous medium, a more complicated pore structure should be discussed to showcase the strength of LBM. An irregular porous medium of porosity 0.4873 and at atmospheric pressure with isothermal condition of 60°C is simulated. Periodic snapshots are organized in Figure 4.7 to describe the invasion patterns in the porous media. Additionally, four various porous media (R1–4) of the same porosity and the particular parameters are simulated to average out the mean evaporation kinetics. At the onset of the evaporation process, a steep fall in the evaporation rate is observed due to the surface breakdown to attain equilibrium meniscus position (Figure 4.7a). Then, the larger openings are prone to invasion due to relatively less capillary pressure, while the smaller openings help to keep the surface wet (Figure 4.7b). As time progresses, the gas phase invades deep inside the porous media and still the smaller openings wet the top of the porous media. In this period, it experiences CRP (Figure 4.7a–e). In this period, it observes maximum fluctuation due to Haines jumps because the overall vapour transport length frequently fluctuates as the meniscus of the smaller openings at the top of the porous media fluctuates with pore invasion deep inside the porous media. After that, even if the smaller openings start to recede, it does not observe a FRP because of the presence of films throughout the solid matrix (see blue lines surrounding the solid matrix) (Figure 4.7a–g). Lastly, the excess liquid clusters in the porous media tend to dry out slowly as the threshold pressure to invade small cluster is high. Therefore, the evaporation rate drops out to observe FRP.

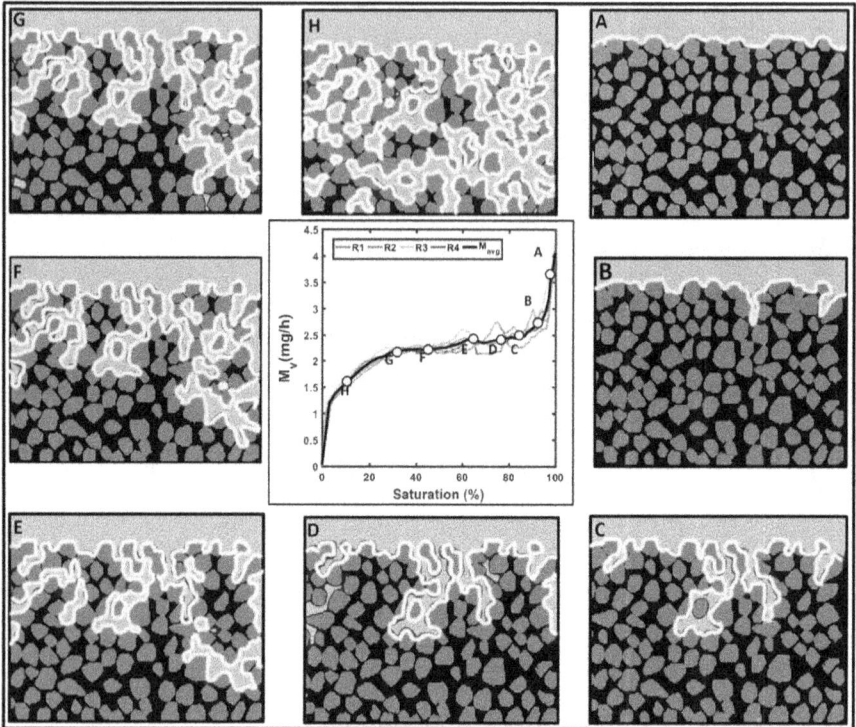

FIGURE 4.7 Micro–macro interaction during evaporation in porous media. (Adapted from Panda et al. (2020b).)

4.3.4 FILM EFFECTS DURING EVAPORATION IN POROUS MEDIA

The evaporation kinetics can be impacted by the formation of liquid films in porous medium. A snapshot of evaporation event is adapted from Figure 4.7c for the discussion in Figure 4.8. A scrutinized observation reveals that there are two kinds of films present in the evaporation phenomenon: (i) film connected to the bulk liquid (in Figure 4.8a) and (ii) thin film surrounding a suspended solid (navigate the arrows in Figure 4.8a and b). Four points are taken in both the cases and the pressure is plotted in Figure 4.8b. In the case of the film connected to the bulk liquid, the pressure (navigate the arrows in Figure 4.8a and b), a pressure gradient is observed which means that it acts like a hydraulic connectivity towards the peak position of the liquid cluster. Such hydraulic connected films are thus termed as "hydraulic" films. On the other hand, the liquid pressure (yellow points) is observed to be almost similar and thus are formed by disrupted hydraulic films. Such films are termed as "adsorbed" films. At the onset of the evaporation process, the hydraulic films dominate over the solid matrix leading to surface wetting for a prolonged period of time. However, as time progresses, the formation of liquid clusters disrupts the hydraulic films and adsorbed films dominate over solitary solid matrix. Due to the wetting characteristic of the water, the adsorbed films are observed on the top which eventually dries out as time progresses.

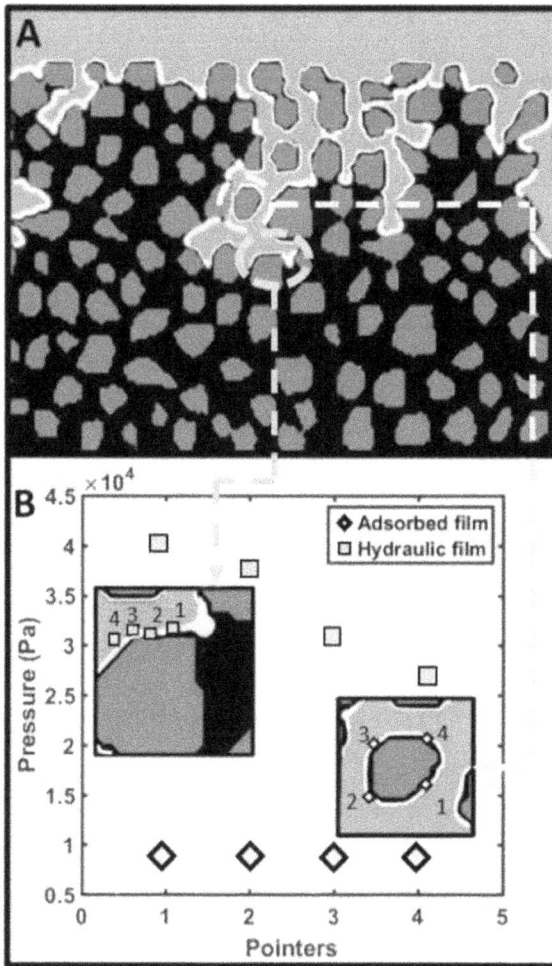

FIGURE 4.8 (a) Representation to exhibit films during evaporation in porous media and (b) liquid pressure variations in hydraulic and adsorbed films. (Adapted from Panda et al. (2020b).)

4.4 CONCLUSION

In this chapter, the physics of isothermal evaporation in capillary porous medium is elucidated using SC LBM simulations. Dominant forces lead to capillary pumping and preferential emptying of liquid in porous medium is explained with LBM simulations on bundle of capillary model. The physics of pore-scale events are presented using the regular pore network with wedged throats and circular pores. These events are (i) an abrupt change in the liquid–gas interface position; (ii) slow invasion, where the interface recedes until the entrance of the pore; (iii) fast invasion in pore to attain the contact angle stability, classified into burst and merge invasions and (iv) Haines jumps, fast invasions cause capillary instabilities which lead to refilling of throats. The micro–macro interactions during evaporation, i.e., pore-scale events' dependencies on evaporation kinetics, are presented using LBM simulations on bundle of capillaries (monomodal and bimodal) and on the random porous media. Complexity of pore-space geometry leads to frequent Haines jumps at the liquid–gas interface and causing more fluctuations in evaporation rates. LBM is a diffusive interface model, and the discretization of the LBM is an essential step to understand the conundrum in the film thickness specification. A quantitative detail about the thickness of the film requires fine mesh and sophisticated collision model. However, it is interesting to note that the results discussed were in a qualitative agreement to different film connectivity.

This work is a step towards the development of complete LBM for evaporation in porous media. Meanwhile, in LBM, it illuminates futuristic simulation capabilities of actual 3D pore structure for which we need to overcome high computational cost. Cross-platform code generation using GPUs and CPUs enable LBM to be utilized in simpler form in various compilers. The locality of the model except the streaming and forcing steps makes the model easier to track the heterogenous boundary complexities in data passing costs. For example, a maximum of three times the length of the domain in the case of D_2Q_9 data movement is expected between two successive cores. Therefore, the ease of parallelization devises LBM as a perfect tool for pore scale modelling. Moreover, the establishment of LBM concepts for evaporation in capillary porous media manifests inherent inclusion of capillary forces, viscous forces and gravity which sunders LBM from classical pore-scale models. The natural ability to confer the interfacial movements incorporates opportunities to establish data-driven empirical models for discrete pore-scale modelling techniques like PNM. For example, the deviation from Hagen Poiseuille due to structural properties can be determined by driving an artificial neural network for finding the permeabilities from the throat cross-sectional images. Thus, it is fascinating to incorporate LBM into PNM to make it more efficient, i.e., by intelligently switching to LBM during capillary unstable regimes. Our perspective of further studies on evaporation in capillary porous media using LBM as a tool includes representation of 3D irregular pore structure and studies of invasion patterns in a large domain. Additionally, micro–macro interactions coupled with thermal effects will extensively be studied in future.

REFERENCES

Abrach, H.E., Dhahri, H., Mhimid, A., 2013. Lattice boltzmann method for modelling heat and mass transfers during drying of deformable porous medium. *J. Porous Media* 16, 837–855. https://doi.org/10.1615/JPorMedia.v16.i9.50.

Al–Ghoul, M., Boon, J.P., Coveney, P. V, Ubertini, S., Succi, S., Bella, G., 2004. Lattice Boltzmann schemes without coordinates. *Philos. Trans. R. Soc. London. Ser. A Math. Phys. Eng. Sci.* 362, 1763–1771. https://doi.org/10.1098/rsta.2004.1413.

Chauvet, F., Duru, P., Geoffroy, S., Prat, M., 2009. Three periods of drying of a single square capillary tube. *Phys. Rev. Lett.* 103, 1–4. https://doi.org/10.1103/PhysRevLett.103.124502.

Chibbaro, S., Falcucci, G., Chiatti, G., Chen, H., Shan, X., Succi, S., 2008. Lattice Boltzmann models for nonideal fluids with arrested phase-separation. *Phys. Rev. E - Stat. Nonlinear, Soft Matter Phys.* 77. https://doi.org/10.1103/PhysRevE.77.036705.

Chin, J., Boek, E.S., Coveney, P. V, 2002. Lattice Boltzmann simulation of the flow of binary immiscible fluids with different viscosities using the Shan-Chen microscopic interaction model. *Philos. Trans. R. Soc. A Math. Phys. Eng. Sci.* 360, 547–558. https://doi.org/10.1098/rsta.2001.0953.

Chopard, B., Hoekstra, A., 2004. Computational science of lattice Boltzmann modelling. *Futur. Gener. Comput. Syst.* 20, 907–908. https://doi.org/10.1016/j.future.2003.12.001.

Colosqui, C.E., Falcucci, G., Ubertini, S., Succi, S., 2012. Mesoscopic simulation of non-ideal fluids with self-tuning of the equation of state. *Soft Matter* 8, 3798–3809. https://doi.org/10.1039/c2sm06353k.

El Abrach, H., Dhahri, H., Mhimid, A., 2013. Numerical simulation of drying of a saturated deformable porous media by the Lattice Boltzmann Method. *Transp. Porous Media* 99, 427–452. https://doi.org/10.1007/s11242-013-0194-2.

Huang, H., Huang, J.-J., Lu, X.-Y., Sukop, M.C., 2013. On simulations of high-density ratio flows using color-gradient multiphase lattice boltzmann models. *Int. J. Mod. Phys. C* 24. https://doi.org/10.1142/S0129183113500216.

Kharaghani, A., Metzger, T., Tsotsas, E., 2012. An irregular pore network model for convective drying and resulting damage of particle aggregates. *Chem. Eng. Sci.* 75, 267–278. https://doi.org/10.1016/j.ces.2012.03.038.

Kupershtokh, A.L., Medvedev, D.A., Karpov, D.I., 2009. On equations of state in a lattice Boltzmann method. *Comput. Math. Appl.* 58, 965–974. https://doi.org/10.1016/j.camwa.2009.02.024.

Lallemand, P., Luo, L.-S., 2000. Theory of the lattice Boltzmann method: Dispersion, dissipation, isotropy, Galilean invariance, and stability. *Phys. Rev. E - Stat. Physics, Plasmas, Fluids, Relat. Interdisc. Top.* 61, 6546–6562. https://doi.org/10.1103/PhysRevE.61.6546.

Laurindo, J.B., Prat, M., 1996. Numerical and experimental network study of evaporation in capillary porous media. Phase distributions. *Chem. Eng. Sci.* 51, 5171–5185. https://doi.org/https://doi.org/10.1016/S0009-2509(96)00341-7.

Lenormand, R., Touboul, E., Zarcone, C., 1988. Numerical models and experiments on immiscible displacements in porous media. *J. Fluid Mech.* 189, 165–187. https://doi.org/10.1017/S0022112088000953.

Liu, H., Kang, Q., Leonardi, C.R., Schmieschek, S., Narváez, A., Jones, B.D., Williams, J.R., Valocchi, A.J., Harting, J., 2016. Multiphase lattice Boltzmann simulations for porous media applications: A review. *Comput. Geosci.* 20, 777–805. https://doi.org/10.1007/s10596-015-9542-3.

Lou, Q., Guo, Z., Shi, B., 2013. Evaluation of outflow boundary conditions for two-phase lattice Boltzmann equation. *Phys. Rev. E - Stat. Nonlinear, Soft Matter Phys.* 87, 1–16. https://doi.org/10.1103/PhysRevE.87.063301.

Martys, N.S., Chen, H., 1996. Simulation of multicomponent fluids in complex three-dimensional geometries by the lattice Boltzmann method. *Phys. Rev. E - Stat. Physics, Plasmas, Fluids, Relat. Interdisc. Top.* 53, 743–750. https://doi.org/10.1103/PhysRevE.53.743.

McCracken, M.E., Abraham, J., 2005. Multiple-relaxation-time lattice-Boltzmann model for multiphase flow. *Phys. Rev. E - Stat. Nonlinear, Soft Matter Phys.* 71. https://doi.org/10.1103/PhysRevE.71.036701.

Metzger, T., Irawan, A., Tsotsas, E., 2007. Isothermal drying of pore networks: Influence of friction for different pore structures. *Dry. Technol.* 25, 49–57. https://doi.org/10.1080/07373930601152640.

Mohamad, A.A., 2011. *Lattice Boltzmann Method: Fundamentals and Engineering Applications with Computer Codes.* SpringerLink : Büche.

Monaco, E., Luo, K.H., Brenner, G., 2011. Multiple relaxation time Lattice Boltzmann simulation of binary droplet collisions. *Lect. Notes Comput. Sci. Eng.* https://doi.org/10.1007/978-3-642-14438-7_27.

Panda, D., Bhaskaran, S., Paliwal, S., Kharaghani, A., Tsotsas, E., Surasani, V.K., 2020a. Pore-scale physics of drying porous media revealed by Lattice Boltzmann simulations. *Dry. Technol.* https://doi.org/10.1080/07373937.2020.1850469.

Panda, D., Supriya, B., Kharaghani, A., Tsotsas, E., Surasani, V.K., 2020b. Lattice Boltzmann simulations for micro-macro interactions during isothermal drying of bundle of capillaries. *Chem. Eng. Sci.* 220. https://doi.org/10.1016/j.ces.2020.115634.

Perumal, D.A., Dass, A.K., 2015. A review on the development of lattice Boltzmann computation of macro fluid flows and heat transfer. *Alexandria Eng. J.* 54, 955–971. https://doi.org/10.1016/j.aej.2015.07.015.

Prat, M., 1992. Drying: percolation theory or continuum approach. *Heat Mass Transf. Porous Med.*

Prat, M., 2007. On the influence of pore shape, contact angle and film flows on drying of capillary porous media. *Int. J. Heat Mass Transf.* 50, 1455–1468. https://doi.org/10.1016/j.ijheatmasstransfer.2006.09.001.

Prat, M., Bouleux, F., 1999. Drying of capillary porous media with a stabilized front in two dimensions. *Phys. Rev. E - Stat. Physics, Plasmas, Fluids, Relat. Interdiscip. Top.* 60, 5647–5656.

Quintard, M., Whitaker, S., 1993. Transport in ordered and disordered porous media: volume-averaged equations, closure problems, and comparison with experiment. *Chem. Eng. Sci.* 48, 2537–2564. https://doi.org/10.1016/0009-2509(93)80266-S.

Sahimi, M., 2011. *Flow and Transport in Porous Media and Fractured Rock: From Classical Methods to Modern Approaches.* Second edition. https://doi.org/10.1002/9783527636693.

Shan, X., Chen, H., 1994. Simulation of nonideal gases and liquid-gas phase transitions by the lattice Boltzmann equation. *Phys. Rev. E* 49, 2941–2948. https://doi.org/10.1103/PhysRevE.49.2941.

Surasani, V. K., Metzger, T., Tsotsas, E., 2008. Consideration of heat transfer in pore network modelling of convective drying. *Int. J. Heat Mass Transf.* 51, 2506–2518. https://doi.org/10.1016/j.ijheatmasstransfer.2007.07.033.

Surasani, V.K., Metzger, T., Tsotsas, E., 2008a. Influence of heating mode on drying behavior of capillary porous media: Pore scale modeling. *Chem. Eng. Sci.* 63. https://doi.org/10.1016/j.ces.2008.07.011.

Surasani, V.K., Metzger, T., Tsotsas, E., 2008b. Consideration of heat transfer in pore network modelling of convective drying. *Int. J. Heat Mass Transf.* 51, 2506–2518. https://doi.org/10.1016/j.ijheatmasstransfer.2007.07.033.

Surasani, V.K., Metzger, T., Tsotsas, E., 2010. Drying simulations of various 3D pore structures by a nonisothermal pore network model. *Dry. Technol.* 28, 615–623. https://doi.org/10.1080/07373931003788676.

Tolman, R.C., 1922. The relation between statistical mechanics and thermodynamics. *J. Am. Chem. Soc.* 44, 75–90. https://doi.org/10.1021/ja01422a009.

Tsotsas, E., Metzger, T., Gnielinski, V., Schlünder, E.-U., 2010. Drying of solid materials. *Ullmann's Encycl. Ind. Chem., Major Reference Works.* https://doi.org/doi:10.1002/14356007.b02_04.pub2.

Wu, R., Kharaghani, A., Tsotsas, E., 2016. Capillary valve effect during slow drying of porous media. *Int. J. Heat Mass Transf.* 94, 81–86. https://doi.org/10.1016/j.ijheatmasstransfer.2015.11.004.

Yiotis, A.G., Boudouvis, A.G., 2004. Effect of liquid films on the drying of porous media 50, 2721–2737. https://doi.org/10.1002/aic.10265.

Yoshida, H., Nagaoka, M., 2010. Multiple-relaxation-time lattice Boltzmann model for the convection and anisotropic diffusion equation. *J. Comput. Phys.* 229, 7774–7795. https://doi.org/10.1016/j.jcp.2010.06.037.

Yuan, P., Schaefer, L., 2006. Equations of state in a lattice Boltzmann model. *Phys. Fluids* 18. https://doi.org/10.1063/1.2187070.

Zachariah, G.T., Panda, D., Surasani, V.K., 2019. Lattice Boltzmann simulations for invasion patterns during drying of capillary porous media. *Chem. Eng. Sci.* 196, 310–323. https://doi.org/10.1016/j.ces.2018.11.003.

Zacharoudiou, I., Boek, E.S., 2016. Capillary filling and Haines jump dynamics using free energy Lattice Boltzmann simulations. *Adv. Water Resour.* 92, 43–56. https://doi.org/10.1016/j.advwatres.2016.03.013.

Zou, Q., He, X., 1997. On pressure and velocity boundary conditions for the lattice Boltzmann BGK model. *Phys. Fluids* 9, 1591–1596.

Zou, Q., Hou, S., Doolen, G.D., 1995. Analytical solutions of the lattice Boltzmann BGK model. *J. Stat. Phys.* 81, 319–334. https://doi.org/10.1007/BF02179981.

5 Pore Network Models for Evaporation in Porous Media

Rui Wu
Shanghai Jiao Tong University

CONTENTS

5.1   INTRODUCTION

In Chapter 4, the Lattice Boltzmann method for evaporation in porous media is introduced. This method is computationally expensive, making the modeling of a REV size porous material prohibitive (REV: representative elementary volume). To disclose the characteristics of evaporation in porous media, which is important for the establishment of the continuum model introduced in the following chapters, it is needed to perform the pore-scale modeling of evaporation in porous media with several REV sizes so as to reveal the link between the transport processes at the pore-scale and the macro-scale. The pore network model has a good trade-off between the description of the interface and pore scale phenomena and the computational efficiency, and is an ideal tool to reveal the multiscale transport processes in porous media and hence has received more and more attention (Metzger, 2019; Prat, 2011). The basis of this model is to approximate the void space of a porous material by a pore network composed of regular pores, i.e., large pore bodies connected by small pore throats. Based on the description of the liquid and vapor transport in each pore in the pore network, the evaporation kinetics can be unraveled. In Chapter 3, we introduce the capillary valve effect, continuous and discontinuous corner films, and capillary instability observed during evaporation in microfluidic pore networks. In this chapter, we will introduce the pore network models depicting the drying processes presented in Chapter 2.

In what follows, the pore network model for evaporation in porous media with the capillary valve effect is introduced. In Section 5.3, the pore network model for evaporation in porous media with the continuous and discontinuous corner films is introduced. The pore network model for capillary instability induced gas–liquid interfaces displacement in porous media during evaporation is introduced in Section 5.4. Finally, the conclusion is drawn in Section 5.5.

5.2   PORE NETWORK MODEL FOR EVAPORATION IN POROUS MEDIA WITH THE CAPILLARY VALVE EFFECT

In this subsection, the pore network model with the capillary valve effect for evaporation in porous media is introduced (Wu et al., 2016). To understand in detail the role of the capillary valve effect, only the slow evaporation case is considered, and the effects of gravity, viscosity, liquid film flow, and heat transfer are neglected. To validate the developed model, the modeling results are compared against the experimental data presented in Section 3.2.

5.2.1   MODEL

To model the slow evaporation processes in the pore network shown in Figure 3.1a, it is needed to determine the invasion threshold pressure of each partially filled pore. The partially filled pore is filled with liquid and has at least one adjacent pore occupied by gas. For each liquid cluster in the pore network during the slow evaporation, the partially filled pore with the lowest invasion threshold pressure will be emptied first. The determination of the invasion threshold pressure for the pore in the pore network shown in Figure 3.1a is presented in Section 3.2.2 and hence is not repeated here.

The vapor transport between two neighboring pores in the pore network during evaporation is depicted as a one-dimensional steady diffusion through stagnant air. The diffusion rate from a pore body to one of its neighboring pore throat with the width of w is

$$Q = \frac{2hP_gD}{RT}ln\left(\frac{P_g - P_{v,pt}}{P_g - P_{v,pb}}\right)/\left(1+\frac{l_{pt}}{w}\right) \tag{5.1}$$

where $h=0.1$ mm is the pore depth (note that all the pores have the same depth); $l_{pt}=1$ mm is the length of the pore throat (all the pore throats have the same length); $R=8.314$ J/mol·K is the gas constant; and D is the diffusivity of vapor, determined as $D=2.14\times10^{-5}[(T+273)/273]2.87$ m^2/s.

In Eq. (5.1), $P_g = 1.014 \times 10^5$ Pa is the total gas pressure, and $P_{v,pb}$ and $P_{v,pt}$ are the vapor pressures in pore bodies and pore throats, respectively. The vapor pressure in a duct is represented by the one at the duct center; and for a partially filled duct, the vapor pressure is equal to the saturated one, which is about 2.984×10^3 Pa at $T = 23.8°C$. Similarly, the diffusion rate between the environment and the pore body attached to the outlet tube with width w is

$$Q = \frac{2hP_gD}{RT} \ln\left(\frac{P_g - P_{v,e}}{P_g - P_{v,pb}} \right) / \left(1 + \frac{2\delta}{w} \right) \tag{5.2}$$

where δ is the length of the outlet tube and $P_{v,e}$ is the vapor pressure in the environment with the relative humidity of 24%.

Now, the slow evaporation processes in the pore network shown in Figure 3.1 can be simulated straightforwardly using the following algorithm: (1) Identify liquid clusters in the pore network. (2) Determine the vapor pressure in each empty duct in the network. (3) Calculate the drying rate of each liquid cluster. (4) Identify the invasion duct of each liquid cluster, i.e., the partially filled duct with the lowest threshold pressure. (5) Calculate the time to empty the invasion duct of each liquid cluster as the volume of liquid in this duct is divided by the drying rate of the cluster, and select the minimum time as the step time. (6) Update the liquid saturation in the invasion duct of each liquid cluster based on the step time determined in step (6). (7) Repeat the previous steps until all liquid in the network is removed.

The above algorithm is very similar to that proposed in Wu et al. (2014), where more details can be found; one difference between them is the determination of the invasion threshold pressure. In the present model, the capillary valve effect (CVE) is taken into account. The detailed explanation of the CVE can be found in Section 3.2.2. In the previous pore network models for evaporation of porous media, however, the CVE is neglected, and the invasion threshold pressures of a partially filled pore are determined as $P_t = 2\sigma \cos\theta_a(1/x + 1/h)$ for which x is the pore width (for a square pore, the width is equal to the length). Here, θ_a is the advancing contact angle.

5.2.2 RESULTS

The optical evaporation experiment based on the quasi-2D PDMS pore network mentioned in Section 3.2.1 is employed to validate the developed model. The evaporation processes in the pore network (i.e., variations of the liquid distribution, number of liquid clusters, and liquid saturation) obtained from the experiment as well as the pore network models with and without the capillary valve effect are compared in Figures 5.1–5.3. As can be seen, the modeling results have a better agreement with the experimental data if the CVE is considered in the pore network model.

If the CVE is not considered in the model, the invasion threshold pressure of a partially filled pore with width x and depth h is determined as $2\sigma \cos\theta_a(1/x + 1/h)$, which implies that for $0 < \theta_a < \pi/2$, pore bodies always have a lower invasion threshold pressure than the connected pore throats, since pore bodies are larger than the adjacent pore throats. Hence, once a pore throat in a liquid cluster is invaded by gas, its neighboring

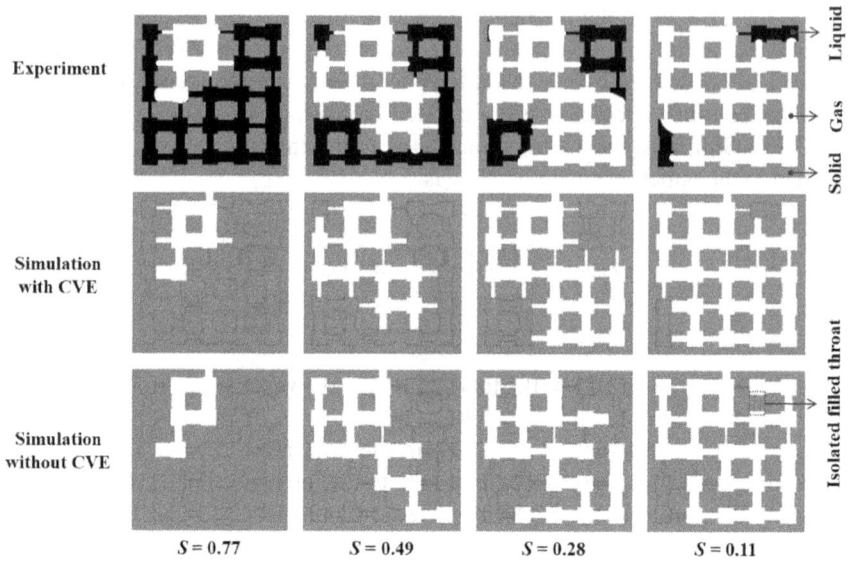

FIGURE 5.1 Variation of the liquid distribution in the pore network with the liquid saturation obtained from the experiment and the pore network model. The liquid saturation is denoted by S.

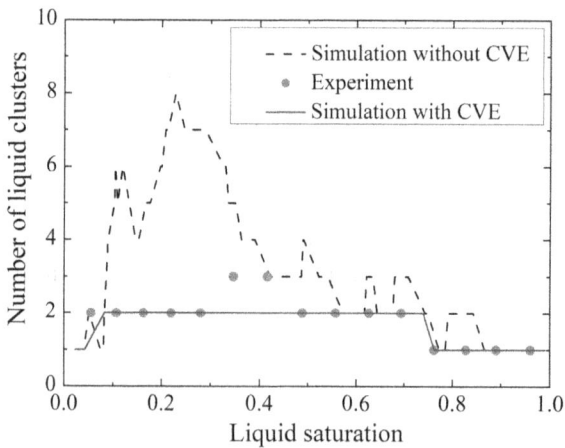

FIGURE 5.2 Variation of the number of the liquid clusters with the liquid saturation.

filled pore body will be emptied immediately in the next step, rendering a "pore throat–pore body" emptying pattern. This in turn results in many isolated filled pore throats in the pore network (Figure 5.1), since the pore throats are more numerous than the pore bodies. In the model with the CVE, the invasion threshold pressure of a pore body can be larger than that of a pore throat. For example, the invasion threshold pressure for bursting invasion into a pore body from a pore throat with width 700 μm is greater than

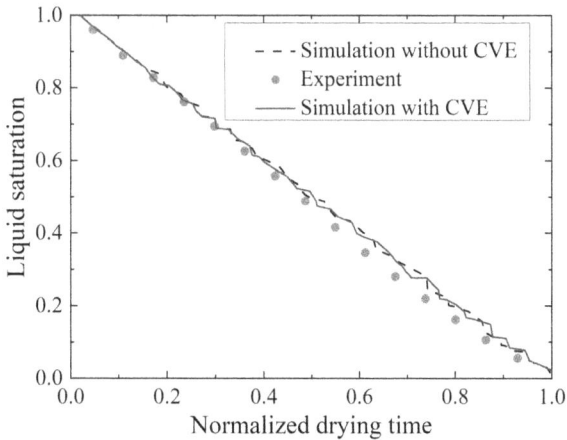

FIGURE 5.3 Variation of the liquid saturation with the normalized evaporation time. The normalized evaporation time is the evaporation time divided by the total evaporation time. The inset shows variation of the liquid saturation with the evaporation time.

the invasion threshold pressure of a pore throat with width 240 µm. As a result, the pore network model without the CVE predicts more isolated filled throats (Figure 5.1) and more liquid clusters (Figure 5.2) as compared to the experiment and the pore network model with the CVE.

As shown in Figure 5.3, variations of the liquid saturation with the normalized evaporation time (i.e., the evaporation time divided by the total evaporation time) obtained from the experiment and the pore network simulations are similar. Nevertheless, the total evaporation times predicted by the pore network models with and without the CVE are about 511 hours, much longer than the experimental data of about 152 hours. This large discrepancy in the total evaporation time could be attributed to the following two reasons. The first is the liquid permeation through Polydimethylsiloxane (PDMS). Although the pore network is covered by glass sheets in our experiment, some liquid in the microfluidic pore network can still be removed through the permeation mechanism. The second reason is liquid film developed along microgrooves at wall surfaces, i.e., the so-called roughness film (Pillai et al., 2009). Note that corner liquid films cannot form in the pores of the PDMS pore network, as mentioned in Section 3.2.1. Pillai et al. (2009) have also revealed a fourfold difference in the total evaporation time between the experiment and the pore network simulation neglecting the roughness films.

Both the liquid permeation through the PDMS pore network and the roughness film flow along the pore walls mentioned above can accelerate the evaporation processes. These two factors, however, are not included in the present pore network models. This is why the total evaporation times predicted by our models are larger than the experimental data. Despite this, the results presented in Figures 5.1 and 5.2 clearly show that the CVE must be taken into account in order to better understand the pore-scale evaporation processes in porous media.

5.3   PORE NETWORK MODEL FOR EVAPORATION IN POROUS MEDIA WITH THE CONTINUOUS AND DISCONTINUOUS CORNER FILMS

For the pore network model introduced in Section 5.2, the corner liquid films are not considered; the viscous effects are also neglected. In this subsection, the pore network model with the corner liquid films is introduced (Wu et al., 2020). The capillary valve effect and the viscous effects are also considered in the model. To validate the developed pore network model, the modeling results are compared against the experimental data presented in Chapter 3.

5.3.1   MODEL

The pore network used in the model is presented in Figure 5.4. To calculate the evaporation rate from the pore network, a cubic boundary zone is introduced right above the pore network surface. The ambient vapor pressure ($P_{v,e}=0$) is applied to all sides of the boundary zone except the side adjacent to the pore network. At the side adjacent to the pore network, the area attached to the outlet pore of the pore network is taken as an inner surface, while the other area is nonpermeable. To evaluate the size of the boundary zone, the vapor pressure at the inner surface between the pore network and the boundary zone is taken as the saturated vapor pressure ($P_{v,s}=9.171\times 10^3$ Pa), and the evaporation rate from this inner surface is calculated. The vapor transport is considered as the steady vapor diffusion through the stagnant air and can be described by

$$\nabla \cdot \left(\frac{DP_g}{RT} \frac{1}{P_g - P_v} \nabla P_v \right) = 0 \tag{5.3}$$

where P_g is the total gas pressure (1.013×10^5 Pa), D the vapor diffusivity ($1.264\times 10^{-5}\,\mathrm{m^2/s}$), R the universal gas constant [8.314 J/mol K], and T the ambient temperature (298 K).

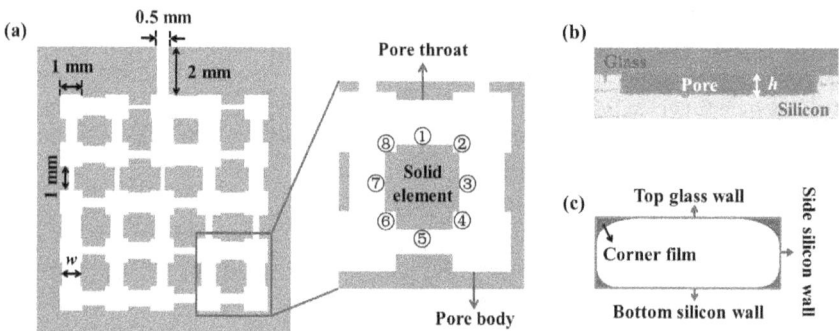

FIGURE 5.4   (a) Schematic of the pore network used in the model. (b) Schematic of corner films in a pore.

The finite volume method is used to solve Eq. (5.3) in the boundary zone. First, the boundary zone is divided into a number of cubic grid cells, each of which has the same side length of 50 μm. Then, Eq. (4.3) is discretized based on the central difference scheme. The variables are stored at the center of each grid cell. Hence, a set of linear equations for the vapor pressure field is generated, which then is solved by the biconjugate gradient stabilized (BiCGSTAB) method. The evaporation rate is determined for various boundary zone sizes. The results indicate that the cubic boundary zone with the side length of 5 mm is enough for determination of the evaporation rate.

In the present pore network model, we assume that the vapor pressure at a filled side wall is equal to the saturated vapor pressure. In previous studies (Yiotis et al., 2004), 1D + 1D model has been used for vapor transport in a pore with corner film. In this 1D + 1D model, the one-dimensional (1D) vapor transfer from the corner film to the center of the pore is considered as the source term in the 1D vapor transport along the centerline of the pore. However, we find that the 1D + 1D model cannot capture the vapor transport in the pores with corner films in the present study. Hence, the 2D model based on Eq. (5.3) is used to describe the vapor transport in the pore network, which, obviously, increases the computational complexity and costs.

We find that if the side walls of the outlet pore and the side walls at edges of the pore network are filled, the vapor inside the pore network is saturated even though the vapor pressure at the entrance of the outlet pore is zero. Since the size of the pore network is not large, our theoretical analysis indicates that there will be continuous corner films at edges of the pore network and at the outlet pore even though only one pore at edges of the pore network is filled. Hence, we assume that the side walls of the outlet pore are filled and that the vapor inside the pore network is saturated. To this end, the evaporation rate from the pore network can be determined by solving Eq. (5.3) in the 3D boundary zone and in the 2D outlet pore by using the finite volume method mentioned above. The outlet pore is divided into a number of square grid cells with the same side length of 50 μm. A grid cell at the entrance of the outlet pore is adjacent to only one grid cell in the boundary zone.

The liquid flow in the pore network is assumed as a fully developed laminar flow. The liquid pressure in a pore, P_l, is calculated at the pore center. The mass flow rate between two adjacent liquid filled pores, e.g., from pore body pj to pore throat pi, is determined as (Ichikawa et al., 2004):

$$Q_{f,pj \to pi} = \frac{16 w_{pi} \rho_l h^3}{\pi^4 \mu_l} \frac{P_{l,pj} - P_{l,pi}}{l_{pj} + l_{pi}} \sum_{n=0}^{\infty} \frac{1}{(1+2n)^4} \left(1 - \frac{\tanh\left[(1+2n)\pi w_{pi} / 2h\right]}{(1+2n)\pi w_{pi} / 2h}\right) \quad (5.4)$$

where ρ_l and μ_l are the density and viscosity of liquid, respectively. If $n=0$ and 1 is used in Eq. (5.4), the calculation error is less than 1% when the aspect ratio, h/w_{pi}, is less than 2. For the pore network in the present study, the aspect ratio is smaller than 0.2. Hence, only $n=0$ and 1 are used in Eq. (5.4).

Each solid element inside the pore network has eight side walls, and a side wall is also the interface between a solid element and a pore, Figure 5.4a. A side wall can be connected to two corner liquid films, Figure 5.4b. We assume that the liquid pressure profiles along the corner films attached to the same side wall are the same. The curvature of corner film in the plane parallel to the liquid flow direction is neglected.

As in previous studies, the mass flow rate along the flow direction of a corner film (e.g., x direction) is determined as (Yiotis et al., 2004):

$$Q_c = -\rho_l \frac{\alpha r_c^4}{\mu_l} \frac{dP_{l,c}}{dx} \tag{5.5}$$

where r_c is the radius of curvature of the corner film in the plane perpendicular to the liquid flow direction and its value depends on the pressure of the liquid in the corner film:

$$r_c = \frac{\sigma}{P_g - P_{l,c}} \tag{5.6}$$

where $\sigma = 0.022 \, \text{N/m}$ is the surface tension.

In Eq. (5.5), α is a dimensionless parameter, which is defined as the ratio of the shape factor to the dimensionless resistance (Dong and Chatzis, 1995). The dimensionless parameter α depends on the geometry of the corner and the contact angle. To determine the value of α, we compute the distribution of liquid velocity in the plane perpendicular to the corner flow direction. The inertial forces are neglected for the liquid flow in the corner films because of low flow velocity. The liquid velocity distribution is computed by COMSOL from which the mass flow rate is obtained. Then, the value of α is determined based on Eq. (5.5) (Wu et al., 2020). The dimensionless parameter α is $\alpha_t = 7.02 \times 10^{-4}$ for the corner film attached to the top wall and $\alpha_b = 9.25 \times 10^{-5}$ for that connected to the bottom wall. The apparent contact angle for the silicon surface is 35° and 0° for the glass surface.

The liquid pressure in the corner films attached to a side wall is determined at the middle point of this side wall in liquid flow direction. The curvature radius of the corner film at a side wall in the plane perpendicular to the flow direction is constant. The mass flow rate from corner films at side wall sj to those at the adjacent side wall si connected to an empty pore is thus determined as

$$Q_{c,sj \to si} = \frac{2}{\mu_l} \frac{\left(\alpha_{t,si} + \alpha_{b,si}\right)\left(\alpha_{t,sj} + \alpha_{b,sj}\right) r_{c,si}^4 r_{c,sj}^4}{l_{c,si}\left(\alpha_{t,si} + \alpha_{b,si}\right) r_{c,si}^4 + l_{c,sj}\left(\alpha_{t,sj} + \alpha_{b,sj}\right) r_{c,sj}^4}\left(P_{l,sj} - P_{l,si}\right) \tag{5.7}$$

where l_c is the length of the corner film, equal to the length of the attached side wall. Here, we consider the liquid flow between two corner films, at least one of which is in an empty pore with no bulk liquid. A filled pore has bulk liquid. The liquid flow between two corner films in filled pores is included in the flow between two adjacent pores, Eq. (5.4). The liquid pressure of the corner films in a filled pore is equal to that of this pore.

During evaporation, the liquid in the pore network is gradually replaced by the gas phase, resembling gas invasion into a liquid filled pore network. This gas invasion process depends on the threshold pressure of each pore and corner. A meniscus can invade a pore (or a corner) only when the pressure difference, $P_g - P_l$, across this meniscus is larger than the threshold pressure of the pore (or the corner). Because of the capillary valve effect, which is induced by the sudden geometrical expansion at the interface between a pore throat and a pore body, the threshold pressure for gas invasion into a filled pore body *pi* from the adjacent pore throat *pj* is determined as

$$P_{t,pi} = \sigma \left(\frac{2\sin\left[\max\left(90°, \theta_{silicon}\right)\right]}{w_{pj}} + \frac{1}{r_h} \right) \tag{5.8}$$

where r_h is the curvature radius of the meniscus in the plane parallel to the height direction, which is the same for all pores:

$$r_h = \frac{h}{1 + \cos\theta_{si}} \tag{5.9}$$

The threshold pressure for gas to invade a filled pore throat *pi* from the adjacent pore body is gained based on the MS-P method (Mayer and Stone, 1965; Princen, 1969):

$$P_{t,pi} = -\sigma \frac{k_1 + k_2 r_c^*}{w_{pi}h - k_3 r_c^{*2}} \tag{5.10a}$$

$$k_1 = (2h + w)\cos\theta_{si} + w \tag{5.10b}$$

$$k_2 = \pi - 3\theta_{si} - (2\cos\theta_{si} - 3\sin\theta_{si} + 2)\cos\theta_{si} \tag{5.10c}$$

$$k_3 = \cos^2\theta_{si} + \cos\theta_{si} - \frac{3}{2}\sin\theta_{si}\cos\theta_{si} - \frac{\pi}{2} + \frac{3\theta_{si}}{2} \tag{5.10d}$$

$$r_c^* = \sqrt{\frac{k_1^2}{4k_2^2} - \frac{wh}{2k_3} + \frac{k_1}{2k_2}} \tag{5.10e}$$

The threshold pressure for gas to invade a corner film attached to side wall *si* is determined as

$$P_{t,si} = \frac{\sigma}{k_c} \tag{5.11}$$

where k_c is the critical radius. We find that the influence of k_c is negligible when k_c is small, e.g., <5 μm. Here, $k_c = 1$ μm is used.

Both the continuous and discontinuous corner films are considered in the present pore network model. The continuous corner films provide liquid flow paths between filled pores. Hence, it is needed to take into account the corner films in order to identify the liquid cluster in the pore network. The procedures for liquid cluster identification are as follows:

(1) Each filled pore and filled solid element have a cluster label, CL, and a cluster number, CN. Initially, CN=0, and CL=0 is applied to all the filled pores and the filled solid elements.

(2) Each filled pore body is scanned. If a filled pore body has a cluster label of $CL=0$, then $CN=CN+1$, and $CL=CN$ is applied to this pore body. Then, we scan each filled pore and each filled solid element; if one has $CL=CN$, we set the adjacent filled pores and filled solid elements with $CL=0$ to have $CL=CN$; this identification is repeated until no filled pores or filled solid elements with $CL=CN$ have connected filled pores or filled solid element with $CL=0$.

(3) Each filled pore throat is scanned. If a filled pore throat has a cluster label of $CL=0$, then $CN=CN+1$, and $CL=CN$ is applied to this pore throat. Then, identification process similar to that in step (2) is performed to determine the liquid cluster composed of filled pore throats and filled solid elements with $CL=CN$. We do not consider the filled pore bodies in this identification process. The reason is that all the pore bodies have been identified or labeled in step (2); hence, in this step, we identify the liquid cluster with pore throats but without pore bodies.

(4) Each filled solid element is scanned. Since all the filled pores have been scanned in steps (2) and (3), if there is a filled solid element with $CL=0$, it must be the isolated one and has no connected filled pores. Hence, if a filled solid element has a cluster label of $CL=0$, then $CN=CN+1$, and $CL=CN$ is applied to this solid element, which is also considered as a liquid cluster. All menisci attached to this isolated filled solid element are set to be moving meniscus.

The following algorithm is used to simulate the evaporation process in the pore network.

(1) Initially, the pore network is fully saturated with liquid except the outlet pore, which is empty but has liquid in its four corners, i.e., corner films. All the menisci are static.

(2) The vapor pressure field in the pore network is determined. The vapor pressure inside the pore network is at the saturated vapor pressure; whereas the vapor pressure in the outlet pore is determined by solving Eq. (5.3).

(3) The evaporation rate from the pore network is calculated. Since the vapor inside the pore network is saturated, the evaporation rate is zero from the saturated side walls and pores inside the pore network. The evaporation rate from a side wall si in the outlet pore is

$$Q_{d,si} = \sum_{i=1}^{m} 2Dh \frac{M_v P_g}{RT} \ln \frac{P_g - P_{v,c,i}}{P_g - P_{v,s}} \qquad (5.12)$$

where $P_{v,c}$ is the vapor pressure in the grid cell and m is the number of grid cells along the length direction of the outlet pore. The evaporation rate from the pore network is equal to the sum of the evaporation rates from the two side walls in the outlet pore.

(4) The liquid clusters in the pore network are identified. Then, the states (moving or static) of menisci in each liquid cluster are checked. If all the menisci in a liquid cluster are static, the meniscus in the filled pore with the lowest threshold is set to be moving. The corners have a larger threshold pressure than pores.

(5) The liquid pressure in filled pores and filled side walls is determined based on the mass conservation law, e.g., for a filled pore pi or a filled side wall si, we have

$$\sum_{pj} Q_{f,pj \to p(s)i} + \sum_{sj} Q_{f,sj \to p(s)i} + \sum_{pk} Q_{d,pk \to p(s)i} = 0 \qquad (5.13)$$

where $Q_{f,pj \to p(s)i}$, $Q_{f,sj \to p(s)i}$, and $Q_{d,pk \to p(s)i}$ are the rates of mass transport into the filled pore pi or the filled side wall si through the flow from the adjacent filled pore pj, through the flow from the adjacent filled side wall sj, and through the diffusion from the adjacent empty pore pk, respectively. Note that the mass diffusion rate inside the pore network is zero because of the saturated vapor.

(6) For each liquid cluster, the pressure difference P_d is determined for each filled pore pi with static menisci:

$$P_d = P_g - P_{l,pi} - P_{t,pi} \qquad (5.14)$$

Then, the filled pores with menisci and positive P_d are scanned among which the static menisci in the pore with the largest P_d are set to be moving.

(7) Repeat steps (5) and (6) until there is no more change of the meniscus state in each liquid cluster.

(8) For each filled pore pi or filled solid element ei with moving menisci, the liquid removal rate is determined as

$$Q_{out,p(e)i} = \sum_{pj} Q_{f,pi \to pj} + \sum_{sj} Q_{f,pi \to sj} + \sum_{pk} Q_{d,p(e)i \to pk} \qquad (5.15)$$

The definition of $Q_{f,pi \to pj}$, $Q_{f,pi \to sj}$, and $Q_{d,pi \to pk}$ can be found in Eq. (5.13). $Q_{d,ei \to pk}$ is the mass diffusion rate from solid element ei to adjacent empty pore pk, which equals to the evaporation rate from the side wall between solid element ei and pore pk. It should be noted that a filled solid element has moving menisci when it is isolated, and the liquid surrounded such solid element can be removed only through the vapor diffusion. Hence, the liquid flow out of the solid element is not considered in Eq. (5.15).

(9) For each filled pore pi or each filled solid element ei with moving menisci, the time to empty the liquid therein is determined:

$$t_{p(e)i} = \frac{\rho_l V_{l,p(e)i}}{Q_{out,p(e)i}} \qquad (5.16)$$

The minimum time t_{min} is selected as the step time. In Eq. (5.16), $V_{l,pi}$ is the volume of the bulk liquid in pore pi and $V_{l,ei}$ is the volume of corner films attached to solid element ei.

The volume of corner films attached to a solid element ei with static menisci (i.e., the solid element is connected to at least one filled pore) is the sum over corner film volumes at the eight side walls of this solid element:

$$V_{l,ei} = \sum_{si}\left(r_{c,si}^2\cos\theta_{si} - \frac{r_{c,si}^2\cos\theta_{si}\sin\theta_{si}}{2} - \frac{\pi - 2\theta_{si}}{4\pi}\pi r_{c,si}^2 \right)l_{si}$$

$$+ \sum_{si}\left(r_{c,si}^2\cos^2\theta_{si} - r_{c,si}^2\cos\theta_{si}\sin\theta_{si} - \frac{\pi - 4\theta_{si}}{4\pi}\pi r_{c,si}^2 \right)l_{si} \tag{5.17}$$

where r_c, determined by Eq. (5.6), is dependent on the pressure of liquid in the corner film. The first and the second term at the right-hand side of Eq. (5.17) are the volumes of the corner films (at the side wall si) attached to the top and bottom walls, respectively. If the filled solid element ei has moving menisci, then the volume of the corner films can be changed during evaporation and is determined by Eq. (5.19).

The volume of bulk liquid in a filled pore pi with static menisci and liquid saturation of one is equal to the difference between the volume of liquid in the pore and the volume of corner films in the pore:

$$V_{l,pi} = V_{pi} - \sum_{si}V_{l,si} \tag{5.18}$$

The first term at the right-hand side of Eq. (5.18) is the volume of pore pi and the second term is the sum of the volumes of corner films in the pore. The volume of bulk liquid in a filled pore with moving menisci is determined by Eq. (5.19).

(10) The volume of liquid in each filled pore pi or filled solid element ei with moving menisci is updated:

$$V_{l,p(e)i} = V_{l,p(e)i} - \frac{t_{min}Q_{out,p(e)i}}{\rho_l} \tag{5.19}$$

(11) Repeat from step (2) until the prescribed condition is reached.

For better understanding of the effects of the corner films, the evaporation processes in the pore network are simulated by not only the above dynamic pore network model with corner films (DPNM_wC) but also the dynamic pore network model without corner films (DPNM_woC). The detailed algorithm of the DPNM_woC can be found in Wu et al. (2019). In addition, the quasi-static PNM with corner films (QSPNM_wC) is also used so as to elucidate the role of liquid viscous forces. In the QSPNM_wC, the liquid viscous forces are not considered, and for each liquid cluster, only the meniscus in the filled pore or filled solid element with the lowest threshold pressure is set to be moving.

5.3.2 RESULTS

To reveal the role of corner films in evaporation in porous media, the simulation results obtained from various PNMs are compared with the experimental data. The evolution of liquid distribution in the pore network during evaporation is shown in Figure 5.5. In the simulation results, the gas, liquid, and solid are shown in white, red, and gray, respectively. In the results obtained from the DPNM_wC, the corner films, attached to solid elements, are also shown in red. From the beginning of evaporation to the total liquid saturation of $S_t = 0.52$, the results obtained from the DPNM_wC and the QSPNM_wC are almost identical. Hence, in Figures 5.5–5.8, the results of the QSPNM_wC are not shown. The difference between the DPNM_wC and the QSPNM_wC will be discussed in detail later.

We do not compare the experimental and simulation results when the total liquid saturation is $S_t < 0.52$. The liquid distribution at $S_t = 0.52$ is shown in Figure 5.5. After this moment, gas invades the second pore body from the left at the bottom of the pore network. By embarking this invasion, the meniscus in the pore body becomes unstable and enters the pores at the lower-left zone of the pore network quickly; at the same time, some empty pores at the center of the pore network are refilled by liquid. This unstable phenomenon is not considered in the pore network model introduced in this subsection. Hence, we compare simulations and measurements for $S_t \geq 0.52$. This unstable phenomenon will be discussed in detail in the next subsection.

During evaporation, liquid is gradually replaced by gas in the pore network, and large liquid clusters are split into smaller ones. At $S_t = 0.82$, an isolated filled solid

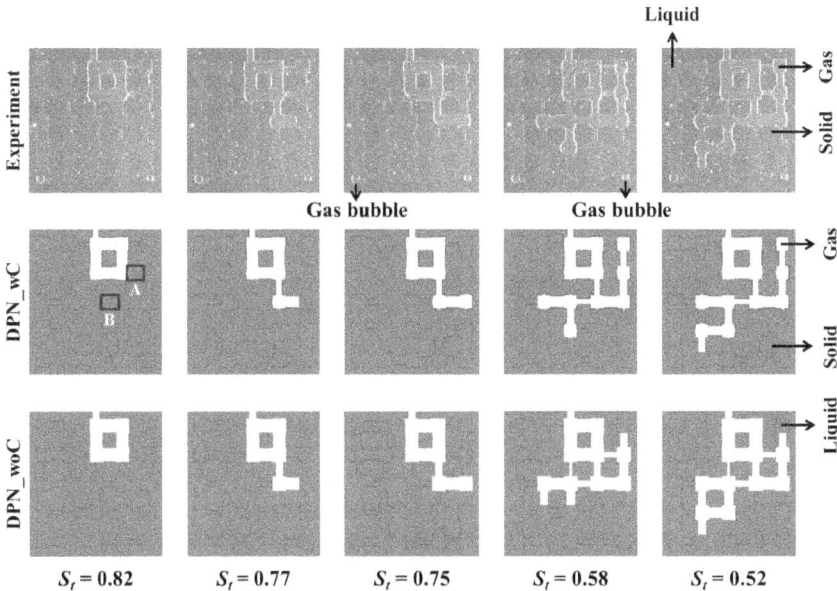

FIGURE 5.5 Variation of liquid distribution in the pore network during evaporation obtained from experiment, DPNM_wC, and DPNM_woC. S_t is the total liquid saturation in the pore network.

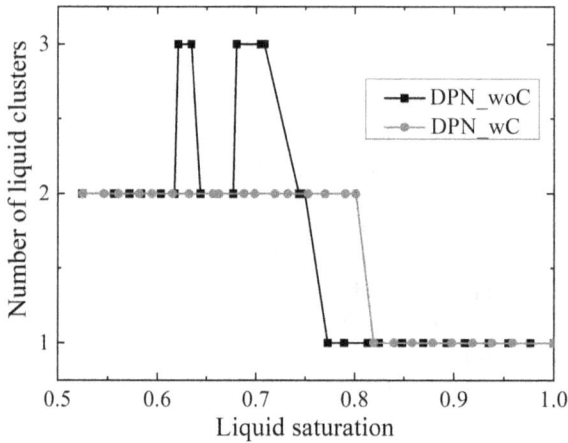

FIGURE 5.6 Variation of the number of liquid clusters during evaporation in the DPNM_wC and the DPNM_woC.

element is observed in the DPNM_wC; the corner films attached to this solid element are discontinuous, since they are not connected to any bulk liquid, Figure 5.5. Since the gas phase is fully saturated with vapor, these discontinuous corner films always exist during evaporation in the DPNM_wC. Experimentally, we also observe that this solid element is covered by liquid. However, in the DPNM_woC, corner films are not considered. Hence, the number of liquid clusters at $S_t=0.82$, as shown in Figure 5.6, is $N_t=2$ in the DPNM_wC, whereas $N_t=1$ in the DPNM_woC.

In the DPNM_woC, two liquid clusters are discerned when gas invades the pore body at the middle of the right edge of the pore network, image of $S_t=0.75$ in Figure 5.5. The liquid cluster at the upper-right zone of the pore network is smaller. The evaporation rates from the large and the small liquid clusters are 6.67×10^{-16} and 3.15×10^{-17} kg/s, respectively. By contrast, in the DPNM_wC, when gas invades the pore body at the middle of the right edge, the number of liquid clusters remains unaltered (Figure 5.6), since the continuous corner films connect the filled pores.

As evaporation proceeds (from $S_t=0.75$ to 0.58), the two pore bodies at the upper-right zone are emptied in the DPNM_wC, consistent with the experimental observations. But in the DPNM_woC, only one pore body is emptied. To understand the difference between these two PNMs, we present in Figure 5.7 the detailed gas invasion process in the upper-right zone of the pore network. In the DPNM_wC, after gas invasion into pore body C (stage II in Figure 5.7), pore throats A and D are still in the same liquid cluster, because they are connected via the continuous corner films. Pore throat D, 520 μm wide, has a lower threshold pressure than pore throat A of 340 μm wide. Hence, pore throat D is invaded in stage III, followed by pore body E in stage IV.

By contrast, after gas invasion into pore body C in the DPNM_woC, filled pore throat A becomes the isolated liquid cluster, stage II in Figure 5.7. Since the volume of liquid in pore throat A is rather small, it is emptied in stage III. Then, pore throat D is invaded in stage IV. Evolution of liquid distribution in stages II and III

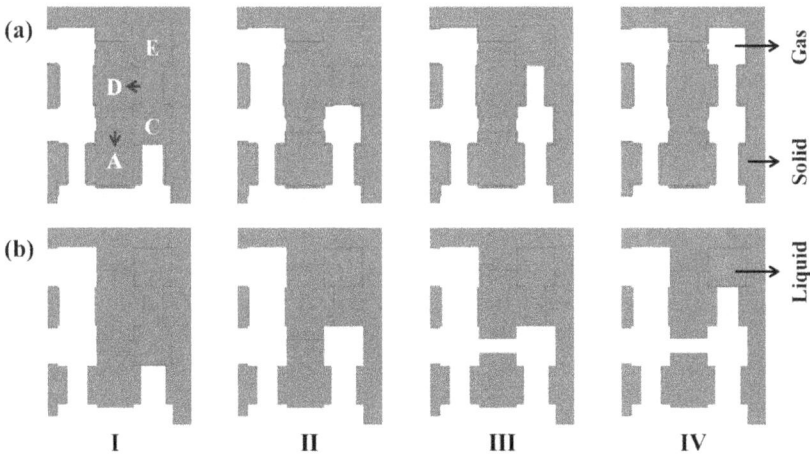

FIGURE 5.7 Gas invasion processes in the upper-right zone of the pore network during evaporation in the (a) DPNM_wC and (b) DPNM_woC.

also explains why the number of liquid clusters fluctuates during evaporation in the DPNM_woC, Figure 5.6.

As evaporation proceeds (from $S_t=0.58$ to 0.52), gas invades the lower-left zone of the pore network. Both the experiment and the DPNM_wC reveal a filled pore throat between two empty pore bodies, image of $S_t=0.52$ in Figure 5.5. However, such filled pore throat is not observed in the DPNM_woC. Moreover, as evaporation continues to $S_t=0.49$, the DPNM_woC shows that the liquid at the upper-right zone is completely removed; whereas both the experiment and the DPNM_wC reveal that the liquid filled pores in this zone are not invaded by gas, since they are smaller and have higher threshold pressure than pores in the lower-left zone.

As gas invades the lower-left zone (as depicted in Figure 5.5), the liquid flow resistance from this zone to the filled pores at the upper-right zone increases, since the length of the corner films increases. To supply the liquid from the lower-left zone to the outlet pore so as to replenish the evaporative loss (which is constant), the liquid pressure in the filled pores at the upper-right zone must decrease (since the liquid flow resistance increases). The filled pore at the upper-right zone is invaded by gas when the liquid pressure in this pore decreases to the value that the pressure difference across the meniscus in this pore, $P_g - P_l$, is equal to or larger than its invasion threshold pressure. Consequently, for evaporation in the pore networks with corner films, the liquid flow induced pressure loss must be taken into account, even for the pore network with small size, e.g., the one in the present study.

As shown in Figure 5.5, evolution of the liquid distribution in the pore network during evaporation can be well predicted by the DPNM_wC, validating its effectiveness. The corner films at the edges of the pore network play an important role in connecting the liquid filled zones in the pore network. The solid elements at the edge of pore network are connected to each other. Hence, corner films at edges of the pore network can be easily connected to bulk liquid if there are filled pores at edges.

In the DPNM_wC, if only one pore at the edges of the pore network is filled, then the corner films at the edge solid elements are the continuous ones and provide the flow paths between the filled zone inside the pore network and the corner films at the outlet pore. This is why corner films always exist in the outlet pore during evaporation in the DPNM_wC, Figure 5.5.

In the DPNM_wC, the evaporation rate from the partially filled pores inside the pore network is zero since the gas phase is fully saturated with vapor in the presence of corner films at the outlet pore. To this end, the generated liquid clusters that are not connected to the outlet pore will not be emptied. As a result, the number of liquid clusters, as shown in Figure 5.6, is always increasing during evaporation. In the DPNM_woC, however, the corner films are not taken into account. The gas phase inside the pore network is not saturated with vapor and this leads to non-zero evaporation rates from the filled pores. As a result, the liquid clusters are generated and removed during evaporation; and hence, the number of liquid clusters is not monotonically increasing, Figure 5.6.

The evolution of the total liquid saturation with the evaporation time obtained by the experiment and PNMs are compared in Figure 5.8. For the experimental results, the total liquid saturation is obtained based on the image analysis. We also conduct an experiment to measure the weight change of the microfluidic pore network by using an electronic balance (Sartorius BT25S, Germany) so as to gain variation of the total liquid saturation during evaporation. The results obtained by these two methods are similar. Since the corner films cannot be discerned in the image analysis, the good agreement between these two methods indicates that the volume of corner films is not significant (but the corner films play an important role in the evaporation processes, as discussed above).

As shown in Figure 5.8, the evaporation process predicted by the DPNM_woC is slower than the experimental result and the one predicted by the DPNM_wC. This is due to the fact that the corner films are not considered in the DPNM_woC. Not

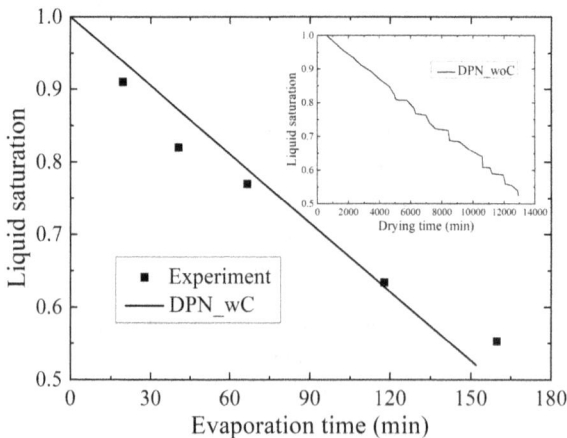

FIGURE 5.8 Variation of the total liquid saturation with the evaporation time.

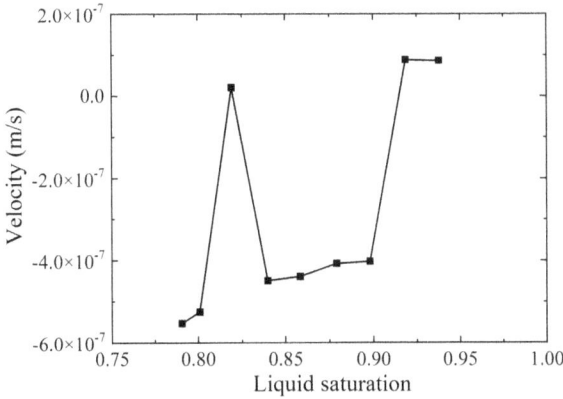

FIGURE 5.9   Variation of the liquid velocity in pore throat B shown in Figure 5.5 during evaporation.

surprisingly, the liquid saturation varies linearly with the evaporation time in the experiment and the DPNM_wC, since the corner films at the outlet pore result in a constant evaporation rate from the pore network as S_t decreases from 1 to 0.52. Interestingly, this linear variation is also observed in DPNM_woC. The reason is that the small pore throats near the outlet pore (Figure 5.5) are filled with liquid during evaporation. As a result, the pore body connected to the outlet pore has almost constant vapor pressure, which in turn leads to an almost constant evaporation rate from the pore network.

As shown in Figure 5.5, the evaporation induced gas invasion in the pore network is a random process. The gas invasion determines the liquid velocity in the pore network. The random gas invasion can result in oscillation of the liquid velocity. To reveal the influence of gas invasion on the liquid velocity field, we present in Figure 5.9 the variation of the averaged liquid velocity, predicted by the DPNM_wC, in the pore throat B shown in Figure 5.5. The flow from the left to the right is defined as the positive direction. As can be seen, the liquid velocity fluctuates during evaporation.

5.4   PORE NETWORK MODEL FOR CAPILLARY INSTABILITY INDUCED GAS–LIQUID INTERFACES DISPLACEMENT IN POROUS MEDIA DURING EVAPORATION

As we mentioned in Chapter 3, during evaporation in porous media, many gas–liquid interfaces can interact with each other, e.g., the refilling of liquid into the pores occupied by gas due to the capillary instability effect. In this subsection, the pore network models for the gas–liquid displacement due to the capillary instability effect are introduced (Zhang et al., 2020). To validate the developed pore network model, the modeling results are compared against the experimental data presented in Chapter 3.

5.4.1 MODEL

In the present pore network model, the inertial forces are considered for flow in all liquid-filled pores. The gas pressure is assumed constant. In addition, the capillary forces, liquid viscous forces, and capillary valve effect are also taken into account in the model.

The liquid flow in a pore is taken as a one-dimensional fully developed laminar flow. A filled pore is the one with the local liquid saturation $s_l > 0$ in pore throats or $s_l > s_{l,re}$ in pore bodies. Here, $s_{l,re}$ is the saturation of the residual liquid. To determine $s_{l,re}$, we assume that the residual liquid in a pore body connects the mouths of two pore throats and its shape is triangle. A partly filled pore ($s_l \leq 1$) contains menisci and has at least one adjacent empty pore, while a fully filled pore ($s_l = 1$) is surrounded by filled pores. The empty pore is the one with $s_l = 0$ for pore throats or $s_l = s_{l,re}$ for pore bodies.

The liquid flow in a filled pore throat k between two filled pore bodies, i and j, can be described as

$$\frac{\partial\left[hw_k \rho_l \left(\frac{l_i}{2} + l_k + \frac{l_j}{2} \right) v_{l,k} \right]}{\partial t} = \left(P_{l,i} - P_{l,j} \right) hw_k - g_k \left(\frac{l_i}{2} + l_k + \frac{l_j}{2} \right) v_{l,k} \quad (5.20)$$

where ρ_l is the liquid density, μ_l the dynamic viscosity, v_l the liquid velocity, and g_k is expressed as (Ichikawa et al., 2004):

$$g_k = \frac{\pi^4 \mu_l}{8\left[1 - \frac{2h}{\pi w_k} \tanh\left(\frac{\pi w_k}{2h} \right) \right]} \frac{w_k}{h}, \quad (5.21)$$

The pore width and length are depicted in Figure 5.4. The liquid flow in a filled pore throat k with $s_{l,k} < 1$ from a filled pore body i to an empty pore body j is depicted as

$$\rho_l hw_k \left\{ \left(\frac{l_i}{2} + s_{l,k}l_k \right) \frac{d^2\left(s_{l,k}l_k \right)}{dt^2} + \left[\frac{d\left(s_{l,k}l_k \right)}{dt} \right]^2 \right\}$$

$$= \left(P_{l,i} - P_{l,k} \right) hw_k - g_k \left(\frac{l_i}{2} + s_{l,k}l_k \right) \frac{d\left(s_{l,k}l_k \right)}{dt}. \quad (5.22)$$

The liquid pressure in a partly filled pore is determined as $P_l = P_g - P_c$ for which the gas pressure P_g is constant. The capillary pressure, P_c, for a moving meniscus in a pore throat i is gained by the MS-P method; see Eq. (5.10). The capillary pressure for a meniscus in a partly filled pore body depends on the shape of meniscus:

$$P_c = \sigma\left(\frac{1}{r_i} + \frac{1}{r_h} \right) \quad (5.23)$$

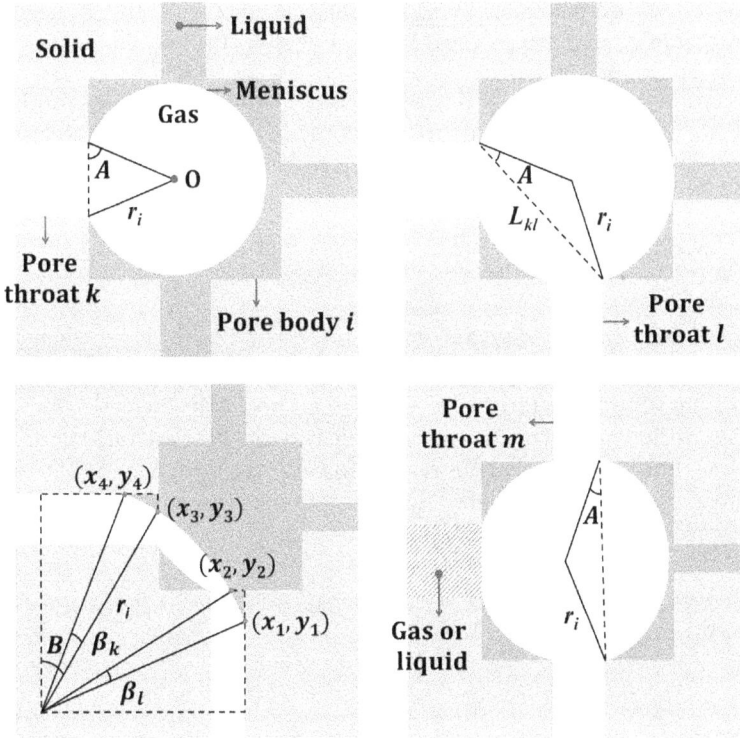

FIGURE 5.10 Schematic of the meniscus configuration in a partially filled pore body i with (a) only one adjacent empty pore throat k; (b) and (c) two adjacent empty pore throats k and l neighboring to each other; and (d) two adjacent empty pore throats l and m opposite to each other.

for which $r_h = h/(1+\cos\theta_s)$, whereas r_i depends on not only the liquid saturation of the pore body but also the state (empty or filled) of the adjacent pore throat. For a partly filled pore body i with only one adjacent empty pore throat (e.g., pore throat k in Figure 5.10a), the curvature radius and the liquid saturation are

$$r_i = \frac{w_k}{2\cos A},$$
(5.24a)

$$s_{l,i} = 1 - \frac{r_i^2\sin(\pi - 2|A|)\frac{A}{|A|} + (\pi + 2A)r_i^2}{2l_i^2},$$
(5.24b)

where A is the angle shown in Figure 5.10a. The contact angle A is positive when the angle between the meniscus and the wall of the pore body is equal to or smaller than 90°, and negative when this angle is larger than 90°.

For a partly filled pore body i with two adjacent empty pore throats neighboring to each other (e.g., pore throats k and l in Figure 5.10b), the curvature radius and the liquid saturation are

$$r_i = \frac{L_{kl}}{2\cos A} \tag{5.25a}$$

$$S_{l,i} = 1 - \frac{L_k L_l}{2l_i^2} - \frac{L_{kl}^2}{4l_i^2}\tan A - \frac{\pi + 2A}{2l_i^2}r_i^2 + \frac{1}{8l_i^2}(l_p - W_l)(l_p - W_k) \tag{5.25b}$$

where $L_k = W_k/2 + l_p/2$, $L_l = W_l/2 + l_i/2$, $L_{kl} = \sqrt{L_k^2 + L_l^2}$, and A the contact angle shown in Figure 5.10b. In Figure 5.10b, the meniscus in the pore body is connected to the mouths of two pore throats. As liquid in the pore body increases, the shape of the meniscus changes. When the contact angle between the gas–solid interface of an empty pore throat and the meniscus reaches to $\pi - \theta_s$, the meniscus starts to enter this pore throat, as illustrated in Figure 5.10c. If meniscus first enters pore throat l, the curvature radius and the liquid saturation are

$$r_i = \frac{W_l + l_p}{2\left[\cos\theta_{s,l} + \cos(2A - \theta_{s,l})\right]} \tag{5.26a}$$

$$S_{l,i} = \frac{I_p - I_{t,l}}{l_p^2}, \tag{5.26b}$$

$$I_p = \frac{1}{2}\left[r_i\cos\theta_{s,l} + r_i\cos(2A - \theta_{s,l})\right]\left[r_i\sin(2A - \theta_{s,l}) - r_i\sin\theta_{s,l}\right]$$

$$- \frac{1}{2}\left[(\pi - 2A)r_i^2 - r_i^2\sin(2A)\right] + \frac{1}{8}(W_l + l_i)(l_i - W_k)$$

$$+ \frac{1}{2}l_i(l_i - W_l) + \frac{1}{8}(l_i - W_l)(l_i - W_k) \tag{5.26c}$$

$$I_{t,l} = \frac{r_i^2\sin\beta_l}{2} + \frac{|x_1 - x_2||y_1 - y_2|}{2} - \frac{\beta_l r_i^2}{2} \tag{5.26d}$$

$$(x_1, \ y_1) = (r_i\cos\theta_{s,l}, \ r_i\sin\theta_{s,l}) \tag{5.26e}$$

$$(x_2, \ y_2) = \left(\sqrt{r_i^2 - y_2^2}, \ r_i\sin(2A - \theta_{s,l}) - (l_i + W_k)/2\right) \tag{5.26f}$$

$$\beta_l = \arccos\left[1 - \frac{(x_1 - x_2)^2}{2r_i^2} - \frac{(y_1 - y_2)^2}{2r_i^2}\right] \tag{5.26g}$$

When meniscus enters both pore throats l and k, the curvature radius and the liquid saturation are

$$r_i = \frac{W_l + l_i}{2\left[\cos\theta_{s,l} + \cos B\right]}, \tag{5.27a}$$

$$S_{l,i} = \frac{I_p - I_{t,l} - I_{t,k}}{l_i^2} \tag{5.27b}$$

$$
\begin{aligned}
I_p = &\frac{1}{2}\left(r_i\cos\theta_{s,l} - r_i\sin\theta_{s,k}\right)\left(r_i\cos\theta_{s,k} - r_i\sin\theta_{s,l}\right) \\
&- \frac{1}{2}\left[\left(\pi - \theta_{s,l} - \theta_{s,k}\right)r_i^2 - r_i^2\sin\left(\theta_{s,l} + \theta_{s,k}\right)\right] \\
&+ \frac{1}{4}\left(W_l + l_i\right)\left(l_i - W_k\right) + \frac{1}{2}l_i\left(l_i - W_l\right) + \frac{1}{8}\left(l_i - W_l\right)\left(l_i - W_k\right)
\end{aligned} \tag{5.27c}
$$

$$I_{t,k} = \frac{r_i^2\sin\beta_k}{2} + \frac{|x_3 - x_4||y_3 - y_4|}{2} - \frac{\beta_k r_i^2}{2}, \tag{5.27d}$$

$$\left(x_3,\ y_3\right) = \left(r_i\cos\theta_{s,l} - \frac{l_i + W_l}{2},\ \sqrt{r_i^2 - x_3^2}\right) \tag{5.27e}$$

$$\left(x_4,\ y_4\right) = \left(r_i\sin\theta_{s,k},\ r_i\cos\theta_{s,k}\right) \tag{5.27f}$$

$$\beta_k = \arccos\left[1 - \frac{\left(x_3 - x_4\right)^2}{2r_i^2} - \frac{\left(y_3 - y_4\right)^2}{2r_i^2}\right] \tag{5.27g}$$

where B is the angle between the vertical line through the center of the meniscus and the line connecting the center and the intersectional point between meniscus and vertical side wall of the pore body, as illustrated in Figure 5.10c. The meniscus will leave the pore body and enters pore throats l and k when the curvature radius of the meniscus reaches

$$r_i = \frac{-b - \sqrt{b^2 - 4ac}}{2a} \tag{5.28a}$$

$$a = 1 - \cos^2\theta_{s,l} - \cos^2\theta_{s,k} \tag{5.28b}$$

$$b = 2W_l\cos\theta_{s,l} + \left(l_i + W_k\right)\cos\theta_{s,k} \tag{5.28c}$$

$$c = -W_l^2 - \left[\frac{\left(l_i + W_k\right)}{2}\right]^2 \tag{5.28d}$$

For a partly filled pore body i with two adjacent empty pore throat opposite to each other (e.g., pore throats l and m in Figure 5.10d),

$$r_i = \frac{A}{|A|} \frac{L_{lm}}{2\cos(A)} \tag{5.29a}$$

$$s_{l,i} = \frac{\left(l_i - \dfrac{W_m + W_l}{2}\right)}{2l_i} - \frac{A}{|A|l_i^2}\left(\frac{(\pi - |2A|)r_i^2}{2} - \frac{r_i^2}{2}\sin|2A|\right)$$
$$+ \frac{(l_i - W_k)(l_i - W_m)}{8l_i^2} + \frac{(l_i - W_l)(l_i - W_k)}{8l_i^2} \tag{5.29b}$$

where contact angle A is positive when the meniscus is concave toward the liquid phase but negative when the meniscus is convex.

Based on the mass conservation law, the following equation is applied to the fully filled pore body:

$$\sum A_j v_{l,j} = 0 \tag{5.30}$$

where A_j is the cross-sectional area of the adjacent pore throats and $v_{l,j}$ is the liquid velocity from the pore throat to the pore body. For a partly filled pore body,

$$\sum A_j v_{l,j} + \frac{dV_{l,i}}{dt} = 0 \tag{5.30}$$

where $V_{l,i}$ is the volume of liquid in the pore body.

The procedure to simulate the capillary instability induced gas–liquid displacement in the pore network is summarized below as well as in Figure 5.11:

(1) The liquid saturation, velocity, and pressure in each filled pore at time t are given.
(2) The volume of liquid, V_l, in each partly filled pore with an active meniscus (i.e., the moving meniscus) at time $t + \Delta t$ is determined. Here, $\Delta t = 1.7 \times 10^{-4}\,\text{s}$ is the step time. The liquid volume of a partially filled pore throat j adjacent to a filled pore body i is updated as

$$V_{l,j}^{t+\Delta t} = hw_j l_j s_{l,j}^{t+\Delta t} \tag{5.31a}$$

$$\left(s_{l,j}^{t+\Delta t}\right)^2 = \frac{2hw_j}{l_j^2 g_j}$$

$$\left\{\left(P_{l,i}^t - P_{l,j}^t\right)\Delta t + \frac{\rho_l}{g_j}\left[\left(P_{l,i}^t - P_{l,j}^t\right)hw_j - g_j s_{l,j}^t l_j v_{l,j}^t\right]\left(e^{\frac{-g_j \Delta t}{\rho_l hw_j}} - 1\right)\right\} + \left(s_{l,j}^t\right)^2 \tag{5.31b}$$

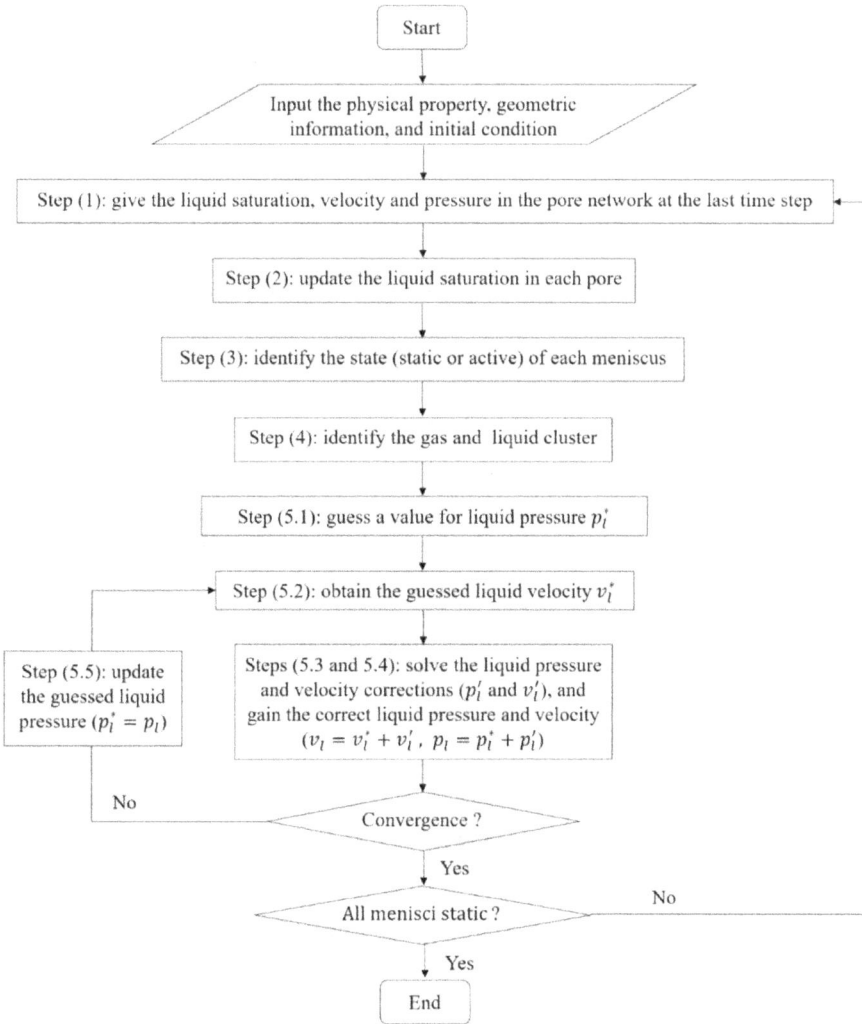

FIGURE 5.11   Flow chart of the pore network modeling algorithm.

If the updated liquid volume in the pore throat j, $V_{l,j}^{t+\Delta t}$, is negative, then the liquid volume in the adjacent filled pore body i is $V_{l,i}^{t+\Delta t} = V_{l,j}^{t+\Delta t} + V_{l,i}^{t}$, and $V_{l,j}^{t+\Delta t}$ is set to be zero. If $V_{l,j}^{t+\Delta t}$ is larger than the volume of the pore throat j, V_j, then the liquid volume in the adjacent empty or partially filled pore body k is $V_{l,k}^{t+\Delta t} = V_{l,k}^{t} + V_{l,j}^{t+\Delta t} - V_j$, and $V_{l,j}^{t+\Delta t}$ is set to b V_j.

The volume of liquid in a partly filled pore body i adjacent to a filled pore throat j is updated as

$$V_{l,i}^{t+\Delta t} = V_{l,i}^{t+\Delta t} + \sum A_j v_{l,j}^{t} \Delta t \tag{5.32}$$

If the updated liquid volume in the pore body i, $V_{l,i}^{t+\Delta t}$, is smaller than the volume of the residual liquid $V_{rel,i}$, then the liquid volume in the adjacent filled pore throat j with the lowest capillary pressure is $V_{l,j}^{t+\Delta t} = V_{l,j}^t - \left(V_{rel,i} - V_{l,i}^{t+\Delta t}\right)$, and $V_{l,i}^{t+\Delta t}$ is set to be $V_{rel,i}$. If $V_{l,i}^{t+\Delta t}$ is larger than the volume of pore body i, V_i, then the liquid volume in each adjacent empty pore throat j is $V_{l,j}^{t+\Delta t} = V_{l,j}^t + \left(V_{l,i}^{t+\Delta t} - V_i\right) \Big/ n_t$, and $V_{l,i}^{t+\Delta t}$ is set to be V_i. Here, n_t is the number of adjacent empty pore throats.

Based on the volume of liquid in each pore, the liquid saturation of each filled pore at time $t + \Delta t$ is calculated straightforwardly.

(3) The state of each meniscus (i.e., the gas–liquid interface) is determined. A static meniscus is the one that cannot move and an active meniscus is movable.

For each partly filled pore throat with an active meniscus and $s_l < 0.1$ or > 0.9, if the liquid flow direction therein is different at time t and $t - \Delta t$, then the meniscus in this partly filled pore throat is labeled as static at time $t + \Delta t$ so as to avoid the numerical error. The explanation is as follows. For instance, as shown in Figure 4.12a, a meniscus in a partly filled pore throat is invading a neighboring empty pore body at time $t - \Delta t$. At time t, the meniscus enters the pore body and the partly filled pore throat becomes fully filled. However, the liquid pressure in the fully filled pore throat can be smaller than the liquid pressure in the pore body. To this end, the meniscus in the pore body may recede to the mouth of the filled pore throat, leading to the change of the direction of the liquid flow in the filled pore throat. But, owing to different curvature radii of the meniscus in the pore body and the pore throat, $P_g - P_l$ across the meniscus can be smaller than the capillary pressure of a moving meniscus in the pore throat. Hence, the meniscus cannot recede into the pore throat and will stop at the mouth of the pore throat. For this reason, this meniscus is set to be static at time $t + \Delta t$. After liquid saturation update at time $t + \Delta t$, the liquid saturation in the pore throat is smaller than 1, but larger than 0.9 (actually close to 1), owing to the small time step.

Similarly, as shown in Figure 5.12b, when a meniscus in the partly filled pore throat invades a filled pore body, it can first enter the pore body and then recede (the liquid flow direction in the pore throat is changed), and finally, stop at the mouth of the pore throat with the liquid saturation being close to 0 (not more than 0.1). As a result, if the liquid flow direction in a partly filled throat with active meniscus and $s_l < 0.1$ (or $s_l > 0.9$) is changed in the last two time steps, the meniscus is set to be static, and the liquid saturation in the pore throat is set to $s_l = 0$ (or $s_l = 1$).

For a partly filled pore throat with $s_l = 1$ and a static meniscus, if $P_g - P_l$ across the meniscus is larger than the capillary pressure of the moving meniscus in the pore throat or smaller than the capillary pressure in the adjacent empty pore body with $s_l = s_{l,re}$, then the static meniscus is set to be active. For a partly filled pore body with $s_l = 1$ and a static meniscus attached to an empty pore throat, if $P_g - P_l$ across the meniscus is smaller than the capillary pressure of the moving meniscus in the connected pore throat or larger than the capillary pressure in the pore body when the contact angle

(a)

Active meniscus   Active meniscus   Static meniscus

Gas

Moving direction   Moving direction

Solid

$t - \Delta t$ t $t + \Delta t$

(b)

Active meniscus   Active meniscus   Static meniscus

Moving direction   Moving direction

$t - \Delta t$ t $t + \Delta t$

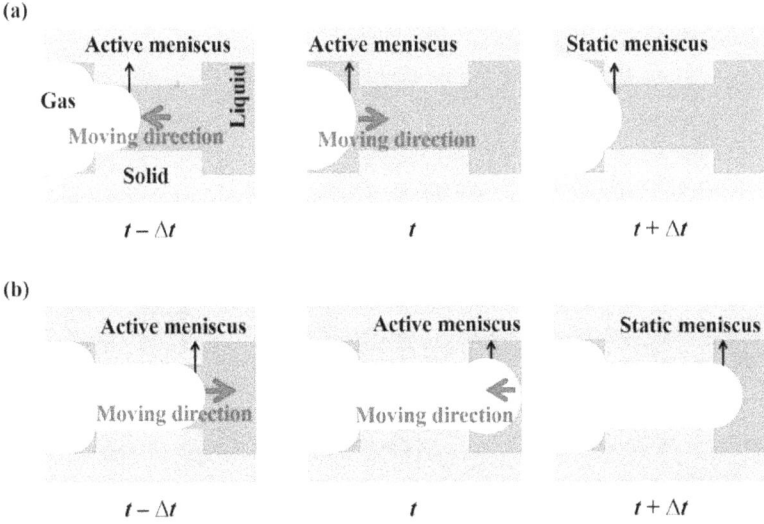

FIGURE 5.12 Schematic of two menisci during transition from active to static state. (a) A meniscus in a pore throat partially filled with liquid invades a neighboring empty pore body. (b) A meniscus in a pore throat partially filled with liquid invades a liquid filled pore body.

between the meniscus and the wall surface is right (the capillary pressure at this moment is the largest because of the capillary valve effect), then the static meniscus is set to be active.

(4) The liquid and gas clusters in the pore network are identified.

(5) The liquid pressure and velocity in each filled pore are determined by the following guess-and-correct method:

(5.1) The pressures of liquid in the partially filled pore with active menisci, $P_l^{t+\Delta t}$, are determined by the capillary pressure, Eqs. (5)–(12). For the partly filled pore throat with static menisci, the liquid velocity, $v_l^{t+\Delta t}$, is zero. The pressures of liquid in the fully filled pore are guessed as $P_l^{*,\ t+\Delta t}$.

(5.2) Based on the guessed liquid pressure field, the guessed liquid velocity in each fully filled pore throat, e.g., pore throat k connecting pore bodies i and j, is determined by solving Eq. (5.20):

$$v_{l,k}^{*,t+\Delta t} = \frac{\left(P_{l,i}^{*,\ t+\Delta t} - P_{l,j}^{*,\ t+\Delta t}\right)\Delta t h w_k + h w_k \rho_l \left(\frac{l_i}{2} + l_k + \frac{l_j}{2}\right)v_{l,k}^t}{h w_k \rho_l \left(\frac{l_i}{2} + l_k + \frac{l_j}{2}\right) + \Delta t g_k \left(\frac{l_i}{2} + l_k + \frac{l_j}{2}\right)}$$
(5.33)

The liquid velocity in the partly filled pore throat with an active meniscus, e.g., pore throat k connected to filled pore body i, is gained by solving Eq. (5.22) analytically:

$$v_{l,k}^{*,t+\Delta t} = \frac{1}{g_k l_k s_{l,k}^{t+\Delta t}}\left\{\left(P_{l,i}^{*,t+\Delta t} - P_{l,k}^{t+\Delta t}\right)h w_k + \left[\left(P_{l,i}^{*,t+\Delta t} - P_{l,k}^{t+\Delta t}\right) - g_k l_k s_{l,k}^t v_{l,k}^t\right]e^{\frac{-g_k \Delta t}{\rho_l h w_k}}\right\}$$
(5.34)

In Eqs. (5.33) and (5.34), if the pore body is the partly filled pore with active menisci, then $P_l^{*,\,t+\Delta t} = P_l^{\,t+\Delta t}$.

(5.3) The correct liquid pressure and velocity are defined as $P_l^{t+\Delta t} = P_l^{*,\,t+\Delta t} + P_l^{',\,t+\Delta t}$ and $v_l^{t+\Delta t} = v_l^{*,\,t+\Delta t} + v_l^{',\,t+\Delta t}$, respectively. Here, P_l' is the liquid pressure correction and v_l' is the liquid velocity correction. It should be noted that the correct liquid pressure and velocity also satisfy Eqs. (5.20) and (5.22), and the following equations can be gained:

$$v_{l,k}^{t+\Delta t} = \frac{\left(P_{l,i}^{t+\Delta t} - P_{l,j}^{t+\Delta t}\right)\Delta t h w_k + h w_k \rho_l \left(\dfrac{l_i}{2} + l_k + \dfrac{l_j}{2}\right)v_{l,k}^t}{h w_k \rho_l \left(\dfrac{l_i}{2} + l_k + \dfrac{l_j}{2}\right) + \Delta t g_k \left(\dfrac{l_i}{2} + l_k + \dfrac{l_j}{2}\right)} \tag{5.35}$$

$$v_{l,k}^{t+\Delta t} = \frac{1}{g_k l_k s_{l,k}^{t+\Delta t}}\left\{\left(P_{l,i}^{t+\Delta t} - P_{l,k}^{t+\Delta t}\right)h w_k + \left[\left(P_{l,i}^{t+\Delta t} - P_{l,k}^{t+\Delta t}\right) - g_k l_k s_{l,k}^t v_{l,k}^t\right]e^{\frac{-g_k \Delta t}{\rho_l h w_k}}\right\} \tag{5.36}$$

Subtracting Eq. (5.33) from Eq. (5.35) yields the liquid velocity correction in the fully filled pore throat:

$$v_{l,k}^{',\,t+\Delta t} = \frac{\Delta t h w_k \left(P_{l,i}^{',\,t+\Delta t} - P_{l,k}^{',\,t+\Delta t}\right)}{h w_k \rho_l \left(\dfrac{l_i}{2} + l_k + \dfrac{l_j}{2}\right) + \Delta t g_k \left(\dfrac{l_i}{2} + l_k + \dfrac{l_j}{2}\right)} \tag{5.37}$$

Subtracting Eq. (5.34) from Eq. (5.36) yields the liquid velocity correction in the partly filled pore throat:

$$v_{l,k}^{',\,t+\Delta t} = \frac{1}{g_k l_k s_{l,k}^{t+\Delta t}}\left(h w_k + e^{\frac{-g_k \Delta t}{\rho_l h w_k}}\right)P_{l,i}^{',\,t+\Delta t}. \tag{5.38}$$

(5.4) Substituting the correct liquid velocity in the filled pore throat, $v_l = v_l' + v_l^*$, into Eq. (4.30) yields

$$\sum A_j\left(v_{l,k}^{',\,t+\Delta t} + v_{l,k}^{*,\,t+\Delta t}\right) = 0 \tag{5.39}$$

Substituting the liquid velocity correction in Eqs. (5.37) and (5.38) into Eq. (5.39) yields a set of linear equations for the liquid pressure correction in each fully filled pore body. For instance, for the fully filled pore body i adjacent to pore body j and filled pore throat k, we have

$$P_{l,i}^{',\,t+\Delta t} = \frac{a_p - Q_p}{a_t} \tag{5.40a}$$

$$a_p = \sum a_{t,k} P_{l,j}^{',\,t+\Delta t} \tag{5.40b}$$

$$a_t = \sum a_{t,k} \tag{5.40c}$$

$$Q_p = \sum A_k v_{l,k}^{*,t+\Delta t} \tag{5.40d}$$

$$A_k = h w_k \tag{5.40e}$$

If the pore throat k is fully filled, $a_{t,k} = A_k^2 \Delta t / \left(h w_k \rho_l l_k s_{l,k}^{t+\Delta t} \right)$; if pore throat k is partly filled with liquid saturation smaller than 1, $a_{t,k} = A_k^2 \big[1 + exp\left(-g_k \Delta t / \rho_l A_k \right) \big] / \left(g_k l_k s_{l,k}^{t+\Delta t} \right)$.

By solving the linear equations, the liquid pressure correction is determined for each fully filled pore body. Then, the liquid velocity correction in each filled pore throat is calculated based on Eqs. (5.37) and (5.38). In this way, the correct liquid pressure, $P_l^{t+\Delta t} = P_l^{*,\,t+\Delta t} + P_l^{',\,t+\Delta t}$, for each fully filled pore body and the correct liquid velocity, $v_l^{t+\Delta t} = v_l^{*,\,t+\Delta t} + v_l^{',\,t+\Delta t}$, for each filled pore throat are obtained.

(5.5) The guessed liquid pressure in each fully filled pore body is updated to equal to the correct liquid pressure gained in step (5.4), i.e., $P_l^{*,\,t+\Delta t} = P_l^{t+\Delta t}$. Repeat from step (5.2) until the prescribed convergence criteria are satisfied.

(6) Repeat steps (5.2)–(5.5) until all menisci become static.

To simulate the observed capillary instability induced gas–liquid displacement in the pore network, the initial liquid pressure at time $t = t_0$ in each filled pore is needed, which, however, cannot be obtained experimentally. To get the initial condition, the following method is employed. Based on the visualization image, we can get the liquid velocity and volume in each filled pore at time $t = t_0 - \Delta t$. The liquid volume in each partially filled pore throat, e.g., pore throat j, at time $t = t_0$ is determined as $V_{l,j}^{t_0} = V_{l,j}^{t_0 - \Delta t} - v_{l,j}^{t_0 - \Delta t} \Delta t$. Then, liquid volume, liquid pressure, and liquid velocity in each filled pore at time $t = t_0$ can be determined based on the algorithm mentioned above and are taken as the initial conditions.

5.4.2 Results

To validate the developed model, we first simulate the refilling of pores with liquid shown in Figure 3.18 in Chapter 3. However, we just focus on the process from stages VI to VII, since the wettability of pore bodies is unknown. Although during this process meniscus A is in pore body 7, the adjacent pore throats 6 and 8 have very similar contact angle (40.5° and 41.7°, respectively), Figure 3.18. Hence, the contact angle of pore body 7 is taken as the averaged value of these two adjacent pore throats. The contact angle of pore throat 2 is 27.5°. The calculated speed of the moving meniscus A agrees well with the experimental data; see the line in Figure 3.19a.

The developed pore network model is also employed to simulate the bubble movement shown in Figure 3.14. The modeling and experimental results are compared in Figure 5.13. Only the top right zone of the pore network (the zone marked by the box

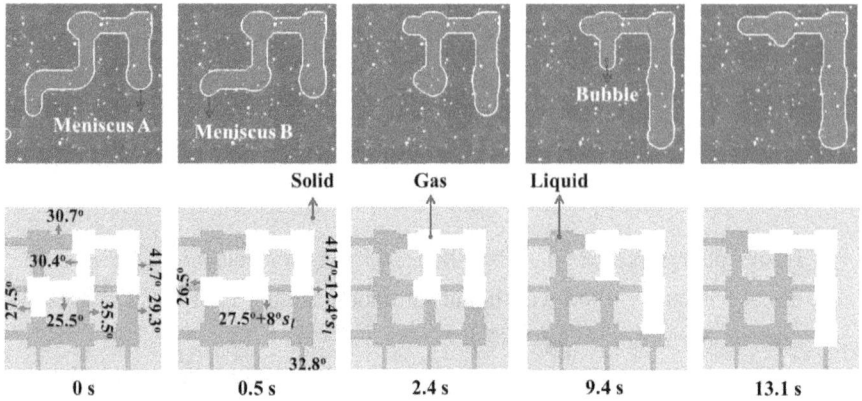

FIGURE 5.13 Bubble movement in the top right zone of the pore network: pore experiment; bottom: pore network model.

in Figure 3.14) is shown in the experimental images. In the modeling images, the solid, liquid, and gas are shown in dark, gray, and white, respectively. In the model, the gas pressure in the bubble is assumed to be uniform and equal to the atmospheric pressure.

The contact angles of pores needed in the model are also shown in Figure 5.13. The contact angles of pore bodies are set based on the contact angles of the neighboring pore throats. It should be noted that the contact angles in the pore throats adjacent to the same pore body can be different, which indicates the non-uniformity of the wettability of the pore body. In the present model, if the maximum difference of contact angles in the pore throats adjacent to the same pore body is more than $7°$, the wettability of the pore body is considered to be non-uniform; then a linear relationship between contact angle θ and liquid saturation s_l is used to describe the wettability of the pore body, Figure 5.13. If the maximum difference of contact angles in the pore throats adjacent to the same pore body is less than $7°$, the contact angle of the pore body is considered to be uniform and is equal to the average of contact angles of the adjacent pore throats, Figure 5.13.

As shown in Figure 5.13, the bubble movement predicted by the pore network model agrees well with the experimental results. It takes a relatively long time for meniscus B to enter the pore body at the center of the pore network (see images of $t=2.4$ and $9.4\,s$). The main reason is that when meniscus B meets the residual liquid in the pore body, its curvature radius, r_i, increases, thereby leading to the reduced capillary pressure and hence the increased liquid pressure, which in turn hinders the movement of meniscus B. The good agreement between the modeling and experimental results in terms of meniscus moving speed (Figure 3.19a) and variation of the liquid phase distribution (Figure 5.13) demonstrates the effectiveness of the developed pore network model.

By using the developed pore network model, we take the bubble movement shown in Figure 5.13 as an example to illustrate the impacts of the inertial forces on the gas–liquid displacement in porous media. To do this, the bubble movement is simulated by the pore network models with the inertial effect (PNMwI) and without the inertial effect (PNMwoI). The surface tension value is varied between $0.001\sigma_f$ and $50\sigma_f$. Because of this variation, the gas bubble can move at different speeds (the higher the surface tension, the faster the bubble moves). Here, $\sigma_f = 0.0221$ N/m^{-1} is the reference surface tension, which is the surface tension of ethanol used in our experiments. In all simulations, the initial conditions are the same, mimicking the experiment (e.g., see Figure 5.13). The average speed of the receding meniscus is calculated from the distance traveled divided by the associated time. The average speed, $v_{r,wi}$, obtained from PNMwI simulations is used to determine the Weber number $(\frac{\rho_l v_{r,wi}^2 h}{\sigma})$ and the capillary number $(\frac{\mu_l v_{r,wi}}{\sigma})$ as well as the Reynolds number $(\frac{\rho_l v_{r,wi} h}{\mu_l})$. The average speed of the receding meniscus is used because more than one advancing meniscus can get involved during the bubble movement (e.g., Figure 5.13).

All the pores in the microfluidic pore network have a rather small height (50 μm). Even if the surface tension is increased by a factor of 50, both the capillary number (1.9×10^{-5}) and the Weber number (1.6×10^{-5}) are rather small. Hence, the capillary forces dominate the bubble movement, and the inertial and viscous forces can be neglected. To this end, the final bubble configuration predicted by the PNMwI and the PNMwoI is almost the same as that in Figure 5.13. The time used to determine the average speed of the receding meniscus is the period from the initial moment to the final stable moment.

In order to elucidate the role of the inertial forces, we change the pore height to 500 μm, and the contact angle values of the silicon surfaces in all pores to 40°. In addition, the widths of pore throats 12, 14, 8, and 10 are changed to 0.4, 0.5, 0.5, and 0.6 mm, respectively (the numbers of pores are illustrated in Figure 5.14). For bubble movement in this transformed pore network, there is only one advancing meniscus (A) and only one receding meniscus (B). We find that the final stable

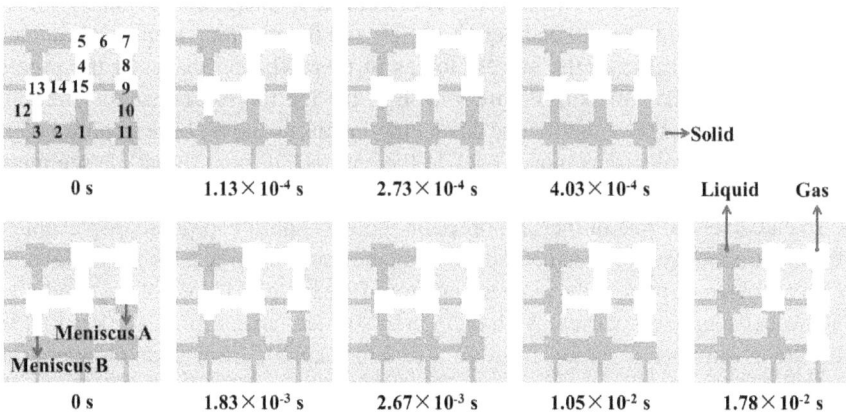

FIGURE 5.14 Evolution of the gas bubble in the transformed pore network at the surface tension of $45.6\sigma_f$ obtained by the pore network model with (top) and without (bottom) the inertial forces.

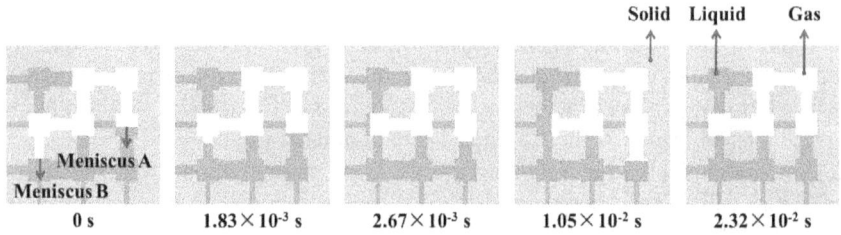

FIGURE 5.15 Evolution of the gas bubble in the transformed pore network at the surface tension of $45\sigma_f$ obtained by the pore network model with the inertial forces.

bubble configuration predicted by the PNMwI and the PNMwoI is different when surface tension is equal to or larger than $46.5\sigma_f$ (the Weber number is 0.085), Figure 5.14. The time steps used in the PNMwoI and PNMwI simulations are 3.33×10^{-8} and 1.66×10^{-6} s, respectively. The average speed of the receding meniscus is the distance from the initial point to the entrance of pore body 13 divided by the time for the receding meniscus moving through this distance. In the PNMwoI, when the advancing meniscus, initially in pore body 9, reaches the entrance of pore throat 10 and the receding meniscus reaches the entrance of the pore body 13, the bubble halts. The reason is that the capillary pressure of the moving meniscus in pore throat 10 is larger than that of pore body 13. However, because of the inertial forces in the PNMwI, the advancing meniscus invades pore throat 10 and pore body 11, and the receding meniscus enters pore body 13 and pore throat 14; see images at time 2.67×10^{-3}, 1.05×10^{-2}, and 1.78×10^{-2} s at the bottom of Figure 5.14.

When the Weber number is smaller than the critical value 0.085, the final bubble configuration in the transformed pore network is the same for PNMwoI and PNMwI. However, we observe the menisci oscillation in PNMwI when the Weber number is smaller than but close to the critical value 0.085 (this menisci oscillation cannot be predicted by the PNMwoI).When the Weber number is 0.082, the advancing meniscus can enter pore throat 10 and pore body 11, and the receding meniscus can enter pore body 13, because of the inertial forces in PNMwI; see images at the time of 2.67×10^{-3} and 1.05×10^{-2} s in Figure 5.15. But, the inertial forces cannot overcome the capillary forces. Eventually, the advancing meniscus moves back to pore body 9 and the receding meniscus moves back to the entrance of pore throat 12; see image at time of 2.32×10^{-2} s in Figure 5.15.

To elucidate the role of inertial forces on the gas–liquid two-phase displacement, we compare the average speed of the receding meniscus moving from the initial point to the entrance of pore body 13 obtained from PNMwI and PNMwoI, $v_{r,wi}$ and $v_{r,woi}$. The Weber number is varied by changing the surface tension from $0.001\sigma_f$ to $45\sigma_f$. The Weber number is varied from 5.1×10^{-5} to 8.2×10^{-2}, Reynolds number from 0.021 to 174.1, and the capillary number from 4.7×10^{-4} to 2.5×10^{-3}. In PNMwoI, the average speed of the receding meniscus is

$$v_{woi} \sim \frac{\sigma}{\mu_l} \qquad (5.41)$$

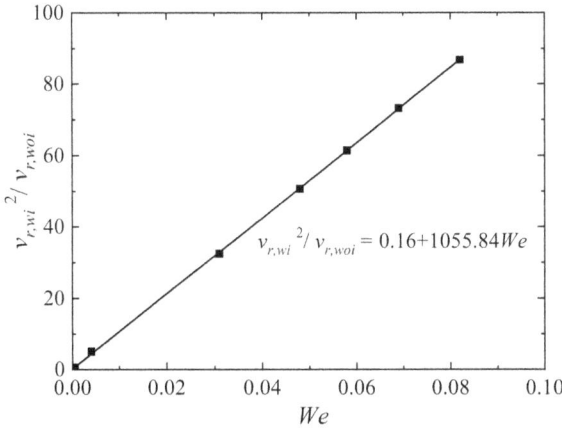

FIGURE 5.16 Variation of $v_{r,wi}^2/v_{r,woi}$ with the Weber number for the gas–liquid displacement in the transformed pore network. $v_{r,wi}$ and $v_{r,woi}$ are the average speed of the receding meniscus obtained from the pore network models with and without the inertial effect, respectively. We is the Weber number.

In PNMwI, the average speed of the receding meniscus can be expressed based on the definition of the Weber number mentioned above:

$$v_{wi} = \sqrt{\frac{\sigma We}{\rho_l h}} , \qquad (5.42)$$

From Eqs. (5.41) and (5.42), we can get

$$\frac{v_{wi}}{v_{woi}} \sim \sqrt{\frac{\sigma We}{\rho_l h}} \bigg/ \frac{\sigma}{\mu_l} = \sqrt{\frac{\mu_l We}{\rho_l h \sigma}} = \sqrt{\frac{We}{\rho_l h v_{woi}}} \qquad (5.43)$$

Eq. (5.43) indicates that

$$\frac{v_{wi}^2}{v_{woi}} \sim We \qquad (5.44)$$

Eq. (5.44) shows that v_{wi}^2/v_{woi} is a linear function of We, i.e., $v_{wi}^2/v_{woi} = c_1 + c_2 We$, which agrees well with the pore network simulation results shown in Figure 5.16. Our calculations show that the coefficients c_1 and c_2 depend on the pore network structure and wettability as well as on the physical properties of the liquid (e.g., density and viscosity).

5.5 SUMMARY

Pore network models for evaporation in porous media considering the capillary valve effect, continuous and discontinuous corner liquid films, and capillary instability effect are presented. The developed pore network models are validated by comparing the modeling results against the experimental data presented in Chapter 3. If the capillary valve effect is not considered in the pore network for drying in porous media without corner liquid films, then more isolated filled pore throats and more liquid clusters are predicted as compared to the experimental data. By comparing the pore network modeling results and the experimental data for drying in porous media with corner liquid films, we find that the continuous corner liquid films can be interrupted to be the discontinuous ones not only by gas invasion into pores but also by the capillary scissors effect due to the local convex topology of the solid matrix. We find that the ratio of the square of the average meniscus moving speed predicted by the pore network model with inertial effect to the average meniscus moving speed predicted by the model without the inertial effect is a linear function of the Weber number. When the Weber number exceeds a critical value, more pores are invaded by the gas–liquid interface in the pore network model with the inertial effect than in the model neglecting the inertial effect.

REFERENCES

Dong, M., Chatzis, I. 1995. The imbibition and flow of a wetting liquid along the corners of a square capillary tube. *Journal of Colloid and Interface Science* 172: 278–288.

Ichikawa, N., Hosokawa, K., Maeda, R. 2004. Interface motion of capillary-driven in rectangular microchannel. *Journal of Colloid and Interface Science* 280: 155–164.

Mayer, R.P., Stone, R.A. 1965. Mercury porosimetry breakthrough pressure for penetration between packed spheres. *Journal of Colloid and Interface Science* 20: 893–911.

Metzger, T. 2019. A personal view on pore network models in drying technology. *Drying Technology* 37: 497–512.

Pillai, K.M., Prat, M., Marcoux, M. 2009. A study on slow evaporation of liquids in a dual-porosity porous medium using square network model. *International Journal of Heat and Mass Transfer* 52: 1643–1656.

Prat, M. 2011. Pore network models of drying, contact angle and films flow. *Chemical Engineering Technology* 34: 1029–1038.

Princen, H.M. 1969. Capillary phenomena in assemblies of parallel cylinders I. Capillary rise between two cylinders. *Journal of Colloid and Interface Science* 30: 69–75.

Wu, R., Cui, G.M., Chen, R. 2014. Pore network study of slow evaporation in hydrophobic porous media. *International Journal of Heat and Mass Transfer* 68: 310–323.

Wu, R., Kharaghani, A., Tsotsas, E. 2016. Capillary valve effect during slow drying of porous media. *International Journal of Heat and Mass Transfer* 94: 81–86.

Wu, R., Yang, L., Zhao, C.Y. 2019. Evaporation from thin porous media with mixed intermediately-wet and hydrophobic networks. *International Journal of Thermal Sciences* 138: 159–173.

Wu, R., Zhang, T., Ye, C., Chao, C.Y., Tsotsas, E., Kharaghani, A. 2020. Pore network model of evaporation in porous media with continuous and discontinuous corner films. *Physical Review Fluid* 5: 014307.

Yiotis, A.G., Boudouvis, A.G., Stubos, A.K., Tsimpanogiamnnis, I.N., Yortsos, Y.C. 2004. Effect of liquid films on the drying of porous media. *AIChE Journal* 50: 2721–2737.

Zhang, T., Wu, R., Zhao, C.Y., Evangelos, T., Abdolreza, K. 2020. Capillary instability induced gas-liquid displacement in porous media: Experimental observation and pore network model. *Physical Review Fluids* 5: 104305.

6 Continuum Models

Marc Prat

Institut de Mécanique des Fluides de Toulouse

CONTENTS

6.1   INTRODUCTION

Continuum models are by far the most commonly used models of drying in porous media in the applications. This is because alternative models such as pore network models (Prat, 2002) or direct simulations (Panda et al., 2020) can be only used over rather small spatial domains owing to computational limitations. The spatial domains of interest in the applications are generally far larger and can be only considered within the framework of the continuum approach to porous media. This, of course, does not mean that PNM or direct simulations are of no interest in practice. They are quite helpful for understanding the physics of drying, and as exemplified for instance in Attari-Moghaddam et al. (2017b), PNM simulations can be used to determine

DOI: 10.1201/9781003011811-6

continuum model parameters. PNM simulations can be also used to improve continuum models or to validate them (e.g. Attari-Moghaddam et al., 2017b; Talbi and Prat, 2022).

The first continuum model of drying dates back to 1929 with the work of Sherwood (1929). As discussed in some detail in Whitaker (1977), Sherwood model was a simple diffusion equation with a constant diffusion coefficient. However, the transfer mechanisms behind the questionable concept of liquid diffusion were unclear until the central role of capillary effects was identified a few years later (Ceaglske and Hougen, 1937). Then, advances in the continuum models of drying were made by Krischer (1938), Philip and de Vries (1957) and Luikov (1966). However, these models were essentially inferred in an intuitive manner from classical continuum conservation equations. Like for other transport processes in porous media, the situation is safer when one can rely on models derived through rigorous upscaling techniques, such as the volume averaging method (Whitaker 2013) or the homogenization method (Auriault et al., 2010). An early attempt in this direction for drying was made by Whitaker (1977) but at that time the volume averaging method was not fully developed and this attempt cannot be considered as fully satisfactory. Another attempt with the homogenization method (Bouddour et al., 1998) actually avoids the complexity of the two-phase flow upscaling. As discussed in Attari-Moghaddam et al. (2017a), the upscaling approach must combine spatial and time averaging because of the complex dynamics of menisci during the drying process. This aspect is not taken into account in the upscaling techniques developed so far which essentially consider situation in which the menisci are immobile at the scale of the representative elementary volume (e.g. Whitaker, 1986). In other words, the commonly used continuum models of drying still lack a rigorous derivation. Nevertheless, the works of Whitaker (1977) and Bouddour et al. (1998) make clear that the continuum modeling is associated with the concept of length scale separation. According to the concept of representative elementary volume (REV), the length separation concept means that it is possible to identify a small volume over which the parameters of the continuum models can be defined and actually computed (provided that the microstructure is known). The REV size is thus much smaller than the porous domain. According to the upscaling methods, the greater the length scale separation, the better is the accuracy of the continuum model.

In part because of the lack of rigorous derivation, there is not a single continuum model formulation, even for the simplified situation considered in this chapter where the mass transfer is dominant and heat transfer can be neglected (one can refer to Perré and Turner (1999) or Vu and Tsotsas (2018) and references therein for models considering the heat transfer). Various formulations have been proposed which can be first classified depending on the class of porous materials. In this respect, one can distinguish the capillary porous media with pores typically greater than or equal to about 1 μm from the hygroscopic porous media with pores in the submicronic range. The main simplification with this classification is that adsorption and Kelvin effect (Adamson, 1990) and the bound water flow (Chen and Pei, 1989) can be a priori neglected in the modeling of capillary porous media. In this chapter, the focus is on capillary porous media to be consistent with the other chapters in the book. Nevertheless, the modeling of hygroscopic porous media will be occasionally

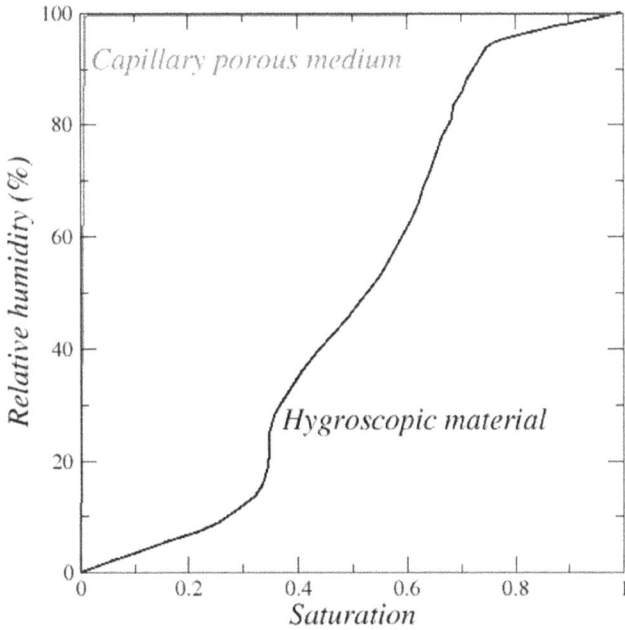

FIGURE 6.1 Desorption isotherms for capillary porous media (schematic) and a hygroscopic porous medium (concrete). (Adapted from Wu et al. (2015).)

evoked. Then, one can distinguish the continuum models relying on the so-called local equilibrium (LE) assumption from the non-local equilibrium (NLE) models. Consider a small sample of porous in an enclosure containing air at a given relative humidity RH_{inf}. Then, by varying step by step the relative humidity (from about 100% to low values) and measuring the weight of the sample, one can measure the desorption isotherm, i.e. the saturation in the sample as a function of the relative humidity. It corresponds to a series of equilibria between the porous medium and the surrounding humid air in the enclosure. The desorption isotherm is denoted by $\varphi(S) = P_{veq.}/P_{vsat.}$, where $P_{vsat.}$ is the saturated vapor pressure and $P_{veq.}$ is the equilibrium vapor partial pressure corresponding to liquid saturation S (S is the volume fraction of the pore space occupied by the liquid). Typical shapes of $\varphi(S)$ are shown in Figure 6.1. As illustrated in Figure 6.1, in a purely capillary porous media, $\varphi(S) = 1$, i.e. $P_{veq.} = P_{vsat.}$ when $S > 0$. The desorption isotherm is sometimes expressed simply as $\varphi(S) = \exp\left(-\dfrac{P_c\, M_v}{\rho_l RT}\right)$ (e.g. Li et al., 2019), which is a macroscopic version of Kelvin relationship (Adamson 1990). However, this is a simplification since liquid adsorption on the pore walls can contribute to the water fixation in addition to capillary effects.

In the LE models, it is assumed that the equilibrium relationship $\varphi(S)$ can be used at the REV scale in drying although drying is a non-equilibrium process. The length scale separation argument can be invoked to justify the use of the desorption isotherm, i.e. an equilibrium relationship, to model a non-equilibrium process. Nevertheless, this approach has been questioned (Benet and Jouanna, 1982; Ouedraogo, et al., 2013) and NLE models have been proposed for both hygroscopic (Lozano et al., 2008; Ouedraogo, et al., 2013) and capillary porous media (Ahmad et al., 2020). The NLE assumption is supported by PNM simulations as reported in Attari-Moghaddam et al. (2017b) for capillary porous media and Maalal et al. (2021) for hygroscopic porous media. Both types of model, i.e. LE and NLE models, will be presented in this chapter.

Drying is essentially a coupled transport problem between the internal transfers within the porous medium and the external transfers in the surrounding gas (humid or dry air in classical drying situations). This means that the full approach to drying requires to solve the equations governing the mass transfer not only in the porous medium but also in the surrounding gas with appropriate interfacial conditions at the porous medium surfaces in contact with the surrounding air. This generally means that the Navier–Stokes equation must be solved in the surrounding air in conjunction with the equation governing the water vapor transport with the additional difficulty that the external flow is generally turbulent. While this type of approach is possible, as reviewed in Defraeye et al. (2012), simplified approaches are generally preferred with the use of the concept of interfacial mass transfer coefficient (e.g. Chen and Pei, 1989; Masmoudi and Prat, 1991) or interfacial resistance (Li et al., 2019). Nevertheless, the boundary condition to be imposed at the evaporative surface is the weakest point in the drying theory. The various options considered in the literature will be discussed in this chapter. In relation with this problem of boundary condition or coupling with the external transfers, a major question in the drying theory is whether the theory is fully predictive or not. "Fully predictive" means that the theory is free of adjustable parameters. In this respect, as we shall see, the drying models based on the continuum approach to porous media cannot be considered as fully predictive.

For convenience, the liquid in the porous medium is assumed to be pure water throughout this chapter unless otherwise mentioned but it should be clear that the presented models apply to other volatile pure liquid. Also, for simplicity, we consider the archetypical situation sketched in Figure 6.2 where only the sample top surface is in contact with the external gas (considered as a binary gas formed by air and the vapor of the evaporating species). This surface will be referred to as the top surface or the interfacial surface. Models are of course not limited to the study of this particular configuration.

Drying of capillary porous media is classically described in three main periods (Van Brakel, 1980): the constant rate period (CRP), the falling rate period (FRP) and the receding front period (RFP). A simplified description by soil scientists (e.g. Lehmann et al., 2008; Li et al., 2019) is to consider two main periods or stages: stage 1 and stage 2. Stage 1 corresponds to the CRP and Stage 2 to the FRP and RFP. Stage 1 is characterized by a wet interfacial surface and a constant or nearly constant evaporation rate. Stage 1 is mainly driven by capillary liquid flow to the interfacial surface. During stage 2, the interfacial surface dries out and a receding internal evaporation front forms inside the porous medium. In this chapter, we will often refer to these two main stages.

In what follows, the LE and NLE continuum models are presented in Sections 6.2 and 6.3. The continuum model parameters are discussed in Section 6.4. The various

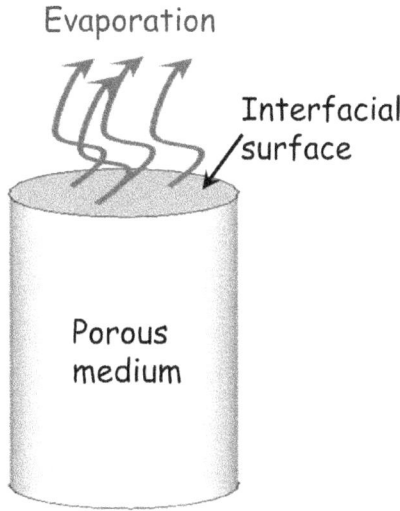

FIGURE 6.2   Archetypical drying situation with interfacial surface at the sample top.

boundary conditions at the interface with the surrounding gas considered in the literature are presented in Section 6.5. Comparisons between the continuum models and experiments are discussed in Section 6.6. Section 6.7 is devoted to a more advanced continuum model where the liquid phase is separated in a percolating liquid phase and a non-percolating liquid phase. A discussion is proposed in Section 6.8. A summary is given in Section 6.9.

6.2   LE CONTINUUM MODELS OF DRYING

6.2.1   LE MODEL NEGLECTING THE VAPOR TRANSPORT

The simplest continuum model of drying considers only the transport in liquid phase due to the capillary action. It is based on the liquid mass conservation equation:

$$\varepsilon \rho_l \frac{\partial S}{\partial t} + \nabla . \left(\rho_l \mathbf{U}_l \right) = 0 \qquad (6.1)$$

where ε, t, S, ρ_l and $\mathbf{U}_l$ denote the porosity, time, liquid saturation, liquid density and liquid filtration velocity respectively. The latter is expressed using the generalized Darcy's law as

$$\mathbf{U}_l = -\frac{k k_r}{\mu} \nabla P_l \qquad (6.2)$$

where P_l is the pressure in the liquid phase, k is the medium permeability, k_r is the liquid phase relative permeability and μ is the liquid viscosity. By introducing the

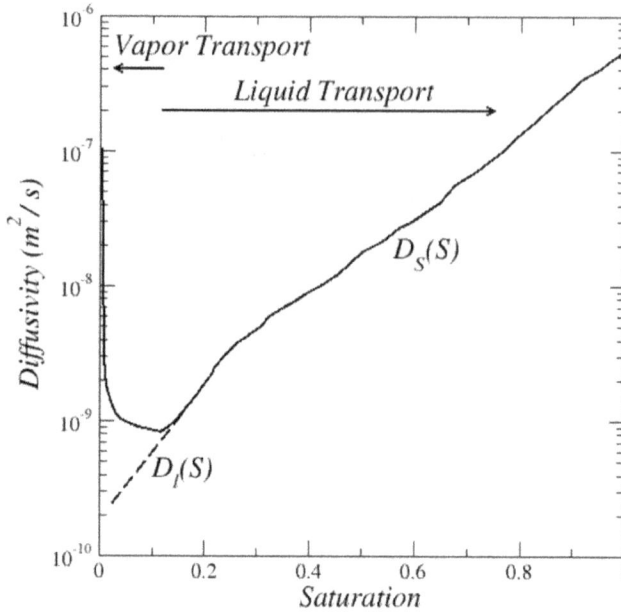

FIGURE 6.3 Typical variations of liquid and total diffusivities as a function of saturation. (Adapted from Pel et al. (1993, 2002).)

capillary pressure curve, $P_c(S)$, where $P_c(S)$ is the local difference between the pressure in the gas phase, assumed a constant equal to the atmospheric pressure throughout this chapter, and the pressure in the liquid phase, Eq. (6.1) is expressed as

$$\varepsilon \rho_l \frac{\partial S}{\partial t} = \nabla \cdot \left(\rho_l D_l(S) \nabla S \right) \tag{6.3}$$

where

$$D_l(S) = -\frac{k k_r}{\mu} \frac{dP_c}{dS} \tag{6.4}$$

Owing to the non-linear dependence of P_c and k_r with S (Dullien, 1992), $D_l(S)$ is a strongly non-linear function of S. This is illustrated in Figure 6.3.

6.2.2 LE MODEL WITH VAPOR TRANSPORT

To establish the LE Model with vapor transport, the liquid mass conservation equation is expressed as

$$\varepsilon \rho_l \frac{\partial S}{\partial t} = \nabla \cdot \left(\rho_l D_l(S) \nabla S \right) - \dot{m} \tag{6.5}$$

where $\dot{m}$ is the phase change rate between the liquid and vapor phases at the REV scale. The water vapor mass conservation is expressed as

$$\nabla.\left(\varepsilon\beta(1-S)D_{eff}\frac{M_v}{RT}\nabla P_v \right) + \dot{m} = 0. \tag{6.6}$$

where D_{eff} is the vapor effective diffusion coefficient. The latter is often expressed as $D_{eff} = \tau D_{vm}$, where D_{vm} is the vapor molecular diffusion coefficient and τ is the tortuosity. A classical relationship (Millington and Quirk, 1961) reads $\tau = \varepsilon^{\frac{1}{3}}(1-S)^{7/3}$. It can be first noticed that Eq. (6.6) is based on a quasi-steady state assumption. This is classical assumption based on the fact that the characteristic time of vapor diffusion is small compared to the drying time. β is the enhancement factor introduced by Philip and de Vries (1957). They postulated that diffusion in the vapor phase in the presence of a temperature gradient would be enhanced due to the presence of liquid islands. The proposed enhancement would be the result of condensation at the warmer end of the liquid island followed by evaporation at the cooler end. The end result is that the liquid islands provide a path of reduced resistance that short circuits the normal vapor phase diffusive path which must circumvent the liquid islands. This is quite different from the classical models for vapor diffusion where the liquid islands are obstacles to the vapor phase diffusion path and, thus, lead to a decrease in the vapor phase diffusion coefficient with increasing saturation. Accordingly, there should be no enhancement when the temperature is spatially uniform as considered throughout this chapter. However, the enhancement factor is often considered even under isothermal conditions to obtain a reasonable agreement between experiment and simulations at low saturations (e.g. Shokri et al., 2009 and references therein). Under isothermal conditions, the enhanced vapor transport implies variation in the meniscus curvature between both ends of the liquid island and impact of the curvature on the equilibrium vapor pressure at the menisci (e.g. Shahraeeni and Or, 2012a). This is possible in sub-micronic pores where the Kelvin effect can have an impact but a priori not in the larger pores of capillary porous media. As pointed out in Shokri et al. (2009), the enhancement transport observed in some experiments might be due to the flow in corner films and not to an enhanced vapor transport. The conclusion is that no vapor transport enhancement effect should be considered in the case of the isothermal drying in capillary porous media. Hence, $\beta = 1$ in Eq. (6.6).

Then, Eqs. (6.5) and (6.6) are added to obtain

$$\varepsilon\rho_l\frac{\partial S}{\partial t} + \nabla.\left(\rho_l D_l(S)\nabla S - \varepsilon\beta(1-S)D_{eff}\frac{M_v}{RT}\nabla P_v \right) = 0 \tag{6.7}$$

Then, the LE assumption is made assuming that the desorption isotherm relationship $\varphi(S)$ is still valid at the REV scale. This leads to express Eq. (6.7) as

$$\varepsilon\rho_l\frac{\partial S}{\partial t} + \nabla.\left(\rho_l(D_l(S) + D_v(S))\nabla S \right) = 0 \tag{6.8}$$

where

$$D_v(S) = -\varepsilon\beta(1-S)D_{eff}\frac{M_v}{\rho_l RT}P_{vsat}\frac{\partial\varphi}{\partial S} \tag{6.9}$$

is the vapor diffusivity. Defining the total diffusivity as $D_S(S) = D_l(S) + D_v(S)$, the LE model is finally expressed as

$$\varepsilon\rho_l\frac{\partial S}{\partial t} = \nabla.\left(\rho_l D_S(S)\nabla S\right) \tag{6.10}$$

Typical variation of D_S with S is illustrated in Figure 6.2. As can be seen, D_S varies on several orders of magnitude, which makes Eq. (6.10) a strongly non-linear diffusion equation.

6.2.3 NLE CONTINUUM MODEL OF DRYING

The NLE continuum model is a two-equation model based on Eqs. (6.5) and (6.6), namely

$$\varepsilon\rho_l\frac{\partial S}{\partial t} = \nabla.\left(\rho_l D_l(S)\nabla S\right) - \dot{m} \tag{6.11}$$

$$\nabla.\left(\varepsilon\beta(1-S)D_{eff}\frac{M_v}{RT}\nabla P_v\right) + \dot{m} = 0. \tag{6.12}$$

However, the thermodynamic equilibrium between the liquid and its vapor is not assumed at the REV scale. This means that the average vapor pressure P_v in the REV is different from the equilibrium vapor pressure $P_{veq.}$ when liquid is present in the REV. This leads to express the phase change rate $\dot{m}$ between the liquid and vapor phases, also referred to as the NLE phase change term, as

$$\dot{m} \approx -a_{gl}\frac{M_v}{RT}\eta\left(P_{veq.} - P_v\right) \tag{6.13}$$

where a_{gl} is the specific interfacial area between liquid and gas phases and η is a coefficient. In the case of capillary porous media, $P_{veq.} = P_{vsat.}$ since adsorption and Kelvin effects are neglected. It can be noted that the linear dependence of $\dot{m}$ with the vapor partial pressure difference $\left(P_{veq.} - P_v\right)$ is a first-order linear approximation, presumably valid sufficiently close to equilibrium (Lozano et al., 2008). One can refer to Lozano et al. (2008) for a non-linear formulation farther from equilibrium. To solve the NLE model, two additional parameters must be specified: the parameter η and the specific interfacial area a_{gl}. The latter is a function of saturation. As illustrated in Figure 6.4 for a packing of particles (Culligan et al. 2004) and also shown from pore network simulations (Joekar-Niasar et al., 2008), the variation of a_{gl} with S is typically bell-shaped.

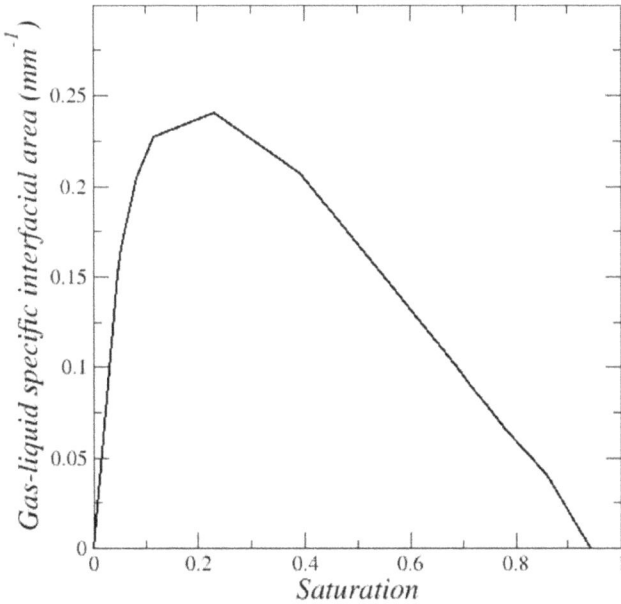

FIGURE 6.4 Typical variation of gas liquid interfacial area a_{gl} with liquid saturation. (Adapted from Culligan et al. (2004).)

Parameter η is a constant not well documented in the literature as discussed in the next section.

6.2.4 METHOD OF SOLUTION

Since the various versions of the continuum model involve parameters varying in a strongly non-linear way with the saturation, the continuum model governing equations and associated initial and boundary conditions are generally solved numerically using standard discretization techniques, such as finite element (Chen and Pei, 1989), finite difference (Kaviany and Mittal, 1987) or finite volume (Hadley, 1985) methods.

6.3 CONTINUUM MODEL PARAMETERS

The use of continuum models implies to specify a series of parameters as input data. In this respect, one can distinguish the internal parameters from the interfacial parameters involved in the top boundary condition. The latter are discussed in Section 6.5. In this section, the focus is on the internal parameters. For the LE model neglecting the vapor transport, the only needed internal parameter is the liquid diffusivity $D_l(S)$, which, as shown by Eq. (6.4), can be expressed as a function of classical parameters such as the permeability, the (drainage) capillary pressure and liquid phase relative permeability curves. These data are available in the literature for a number of porous media (e.g. Dullien, 1992; Kaviany, 1991, among others). When

these data are not available from the literature for a given porous medium, the experimental determination of these parameters or directly $D_l(S)$ is tedious (e.g. Pel et al.,
1993). An alternative is to compute them numerically from digital images of the
microstructure of the porous medium of interest, (e.g. Blunt et al., 2002 and references therein). However, the numerical procedure is not necessarily straightforward
due to possible difficulties to acquire representative image of the microstructures,
to binarize them and to perform the properties' computations (Guibert et al., 2016).
Examples of the numerical approach for simple networks can be found in Attari-
Moghaddam et al. (2017b) and Ahmad et al. (2020). In the case of the LE model with
vapor transport, the vapor diffusivity, $D_v(S)$, must be also determined or directly the
total diffusivity $D_S(S)$ (e.g. Pel et al., 1993). The NLE two-equation model is required
in addition to determine the specific interfacial area a_{gl} (Figure 6.4) and the phase
change coefficient η. Whereas it is possible to determine the former from digital
images (Culligan et al. 2004) or from pore network simulations (Joekar-Niasar et al.,
2008), the situation is less clear as regards η. Within the framework of upscaling
techniques, it is expected that η can be computed on a given microstructure from
a dedicated closure problem (Whitaker 2013). However, such a dedicated closure
problem has not yet been established. So far, the coefficient η in the non-equilibrium
phase change term that has been estimated from experiments (e.g. Lozano et al.,
2008) or from comparisons between experiments and continuum model simulations
(e.g. Li et al., 2019) or from comparisons with pore network simulations (e.g. Ahmad
et al., 2020). It has been much less studied than more classical parameters such as
the capillary pressure curve or the liquid relative permeability. Contrary to these
parameters, there is no mathematical formulation enabling one to predict it, even for
classical model porous media such as a random packing of particles.

6.4   BOUNDARY CONDITIONS AT THE INTERFACIAL SURFACE

To obtain a solution from the continuum models presented in Section 6.3, initial and
boundary conditions must be specified. Referring to Figure 6.2, the controversial issue
is the boundary condition at the top surface, i.e. the top boundary condition (TBC).

6.4.1   TBC WITH CRITICAL SATURATION CONCEPT

As discussed in Li et al. (2019), the simplest approach consists in considering the
boundary condition

$$j = -\rho_l D^* \nabla S . \mathbf{n} = j_{pot} \text{ when } S(z=0) > S_{cr}; \ S(z=0) = S_{cr} \text{ else} \qquad (6.14)$$

where $S(z=0)$ is the saturation at the porous medium top surface; $D^* = D_l(S)$ (LE
model neglecting the vapor transport) or $D^* = D_S(S)$ (LE model with vapor transport); $\mathbf{n}$ is a unit normal vector; j_{pot} is the potential evaporation, i.e. the evaporation
flux when the porous medium is covered by a film of water; S_{cr} is the critical saturation. Thus, in this approach, the evaporation flux is constant and equal to the potential evaporation as long as the saturation at the surface is greater than the so-called

critical saturation. Then, the boundary condition switches to the constant critical saturation boundary condition. As discussed in Li et al. (2019) and Talbi and Prat (2021), this approach is questionable. First, it makes sense only for low evaporation condition when the evaporation rate is approximately constant over stage 1 evaporation. As discussed in Talbi and Prat (2021), the evaporation flux is actually rarely fully constant over stage 1. Furthermore, the evaporation rate during stage 1, even when almost constant, can be lower than the potential evaporation due to the partial invasion of the surface occurring during the so-called breakthrough period (Talbi and Prat, 2021) occurring at the very beginning of drying. A difficulty is to specify the critical saturation S_{cr}. Actually, this parameter is often used as a fitting parameter to obtain a reasonable agreement with the experimental data and not according to a predictive approach.

6.4.2   TBC WITH INTERFACIAL RESISTANCE CONCEPT

An alternative top boundary condition that is also used with Eq. (6.3) or Eq. (6.10) is by introducing an interfacial resistance term to account for the reduction of the evaporation rate when the saturation decreases at the surface of the porous medium. This boundary condition is expressed as

$$j = -\rho_l D^*(S)\nabla S.\mathbf{n} = \frac{M_v}{RT}\frac{\left(P_{vsat} - P_{v\infty}\right)}{r_a + r_{pm}} \tag{6.15}$$

where $P_{v\infty}$ is the vapor partial pressure in the ambience; P_{vsat} is the saturated vapor pressure; M_v, R and T represent the molar mass of water, universal gas constant and temperature. In Eq. (6.15), r_a represents the external boundary layer resistance and r_{pm} is the interfacial resistance due to the mass transfer in the porous medium. In this approach, it is assumed that the external boundary layer resistance r_a can be determined only from the consideration of the external transfer at the beginning of the drying process, i.e. from the potential evaporation expressed as $j_{pot} = \frac{M_v}{RT}\frac{\left(P_{vsat} - P_{v\infty}\right)}{r_a}$. Thus, r_a is considered as fixed for fixed external conditions (surrounding air velocity and relative humidity), whereas r_{pm} increases during the drying process. Formulation of r_{pm} as a function of saturation can be found in Li et al. (2019) and references therein. Compared to the critical saturation approach, Eq. (6.14), the interfacial resistance approach (Eq. 6.15) allows reproducing the decrease in the evaporation rate often observed during stage 1 evaporation and a more gradual transition between stage 1 and stage 2 in better agreement with the observations (Li et al., 2019). The interfacial resistance approach is discussed in detail in Talbi and Prat (2021) from comparisons with PNM simulations. The results presented in Talbi and Prat (2021) indicate that the external interfacial resistance actually varies during drying due to the desaturation of the surface, see also Shahraeeni et al. (2012b). Furthermore, for a given porous medium, the interfacial resistance is not an intrinsic property of the porous medium, i.e. only a function of porous medium variables such as the saturation. It also depends on the external flow properties, such as the external boundary layer thickness.

Another option considered in the literature reads (Li et al., 2019)

$$j = -\rho_l D_{ls}(S) \nabla S . n = \frac{M_v}{RT} \frac{(P_{vi} - P_{v\infty})}{r_a}$$ (6.16)

with $P_{vi} = \varphi(S_{top}) p_{vsat}$. Eq. (6.16) poses the same problems as before. The external resistance should not be considered as a constant for given external condition but varying with the saturation at the top (Talbi and Prat, 2021). Using the bulk desorption isotherm at the top surface is questionable.

6.4.3 TBC WITH MASS TRANSFER COEFFICIENT CONCEPT

In the engineering literature (e.g. Kaviany and Mittal, 1987; Chen and Pei, 1989; Masmoudi and Prat, 1991), the TBC is commonly based on the concept of mass transfer coefficient. This concept originates from classical boundary layer flow models (Schlichting and Gersten, 2016). In the case of a boundary layer flow over a flat surface with a uniform vapor partial pressure P_{vi} over the surface, the evaporation flux along the surface can be expressed as

$$j(x) = h_m(x) \frac{M_v}{RT}(P_{vi} - P_{v\infty})$$ (6.17)

where $h_m(x)$ is the local mass transfer coefficient and x is the position along the plate (with $x=0$ at the beginning of the plate, referred to as the leading edge):

$$h_m(x) = C \frac{D_v}{x} Sc^p Re_x^m$$ (6.18)

where $Sc = \dfrac{\mu_g}{\rho_g D_v}$ is the Schmidt number; $Re_x = \dfrac{\rho_g U_\infty x}{\mu_g}$ is the local Reynolds number; p and m are exponents depending on the nature of the flow, i.e. laminar or turbulent (Kays and Crawford, 1980). Equations (6.17) and (6.18) can be integrated over the plate length to express the mean evaporation flux as

$$j = h_m \frac{M_v}{RT}(P_{vi} - P_{v\infty})$$ (6.19)

where $h_m = \dfrac{1}{L} \displaystyle\int_0^L h_m(x) dx$ and L is the plate length. By analogy, the evaporation at the interfacial surface of the porous medium is expressed as (Chen and Pei, 1989)

$$j = h_m(S) \frac{M_v}{RT}(P_{vi} - P_{v\infty})$$ (6.20)

where P_{vi} is the vapor partial pressure at the top surface of the porous medium, which is generally expressed as $P_{vi} = P_{vsat} \varphi(S)$ assuming that the bulk desorption isotherm relationship is still valid at the surface, which, as mentioned before, is questionable. The new feature compared to classical boundary layer flows is the dependence of the mass transfer coefficient with the saturation. The coefficient decreases with a decreasing saturation at the surface (Kaviany and Mittal, 1987; Chen and Pei, 1989). However, as pointed out in Kaviany and Mittal (1987), there is no theory predicting the variation of $h_m(S)$ with the saturation. In fact, $h_m(S)$ is used as an adjustable parameter in the comparison between the experimental results and the simulations (e.g. Chen and Pei, 1989).

It can be also noted that Eqs. (6.20) and (6.16) are actually similar with $h_m(S) = \dfrac{1}{r_a}$.

6.4.4 TBC with External Mass Transfer Characteristic Length

The TBC can be also expressed as (e.g. Plourde and Prat, 2003)

$$j = \frac{D_{vm}}{\delta} \frac{M_v}{RT}(P_{vi} - P_{v\infty}) \tag{6.21}$$

where δ is the external mass transfer characteristic length scale (the mass transfer boundary layer mean thickness in the case of convective drying for instance). Equation (6.21) is actually very similar to Eqs. (6.16) and (6.20) with $r_a = \dfrac{\delta}{D_{vm}}$ or $h_m = \dfrac{D_{vm}}{\delta}$. The above thus indicates that it is not correct to consider δ as a constant for given external flow conditions since both r_a and h_m must be considered as varying with surface saturation.

6.4.5 TBC for NLE Two-Equation Model

Since the NLE model is a two-equation model, two interfacial boundary conditions must be imposed, one for the liquid transport equation, Eq. (6.11), and one for the vapor transport equation, Eq. (6.12). Consider the situation where both phases, liquid and gas, coexist in the porous medium at the surface. Physically, one then expects that a fraction of the vapor flux at the surface is from the dry pores at the surface, whereas the complementary fraction corresponds to the evaporation from the liquid pores at the surface. Using a boundary layer type expression for the vapor flux at the surface (Section 6.5.4), this is expressed as

$$-\rho_l D_l(S)\nabla S.\mathbf{n} = (1 - A_{surfdry})\, D_v \frac{M_v}{RT} \frac{(P_{vs} - P_{v\infty})}{\delta} \tag{6.22}$$

and

$$-\varepsilon(1 - S)D_{eff}\frac{M_v}{RT}\nabla P_v.\mathbf{n} = A_{surfdry}\, D_{vm} \frac{M_v}{RT} \frac{(P_v - P_{v\infty})}{\delta} \tag{6.23}$$

where $\mathbf{n}$ is the unit normal vector directed from the porous medium toward the external gas boundary layer, δ is the external boundary layer thickness and $P_{v\infty}$ is the vapor partial pressure in the external gas away from the porous medium surface. In Eqs. (6.22) and (6.23), $A_{surfdry}$ represents the fractional contribution of the dry pores to the evaporation rate, whereas $(1.-A_{surfdry})$ represents the fractional contribution of the wet pores. The first choice is to specify $A_{surfdry}$ according to $A_{surfdry} = 1 - S_{surf}$, where S_{surf} is the liquid saturation at the considered surface. However, numerical tests with this choice led to inconsistent results, namely non-monotonous variation of the vapor pressure at the surface in the dry pores. Physically, it is expected that this vapor pressure at the surface gradually decreases along the drying process (Le Bray and Prat, 1999). The following relationship led to consistent results in this respect:

$$A_{surfdry} = \left(1 - \left(\frac{S_{surf} - S_{irr}}{1 - S_{irr}}\right)\right)^q \qquad (6.24)$$

where q is an exponent taken equal to 3 in our simulations and S_{irr} is the irreducible saturation (see the next section for more details). The variation of $A_{surfdry}$ and $A_{surfwet} = 1. - A_{surfdry}$ as a function of the reduced saturation at the surface $S^* = \left(\frac{S_{surf} - S_{irr}}{1 - S_{irr}}\right)$ computed using Eq. (6.24) is shown in Figure 6.5. The evolution in Figure 6.5 reflects the fact that the relative contribution of the dry pores at the surface to the evaporation rate increases with the decreasing surface saturation.

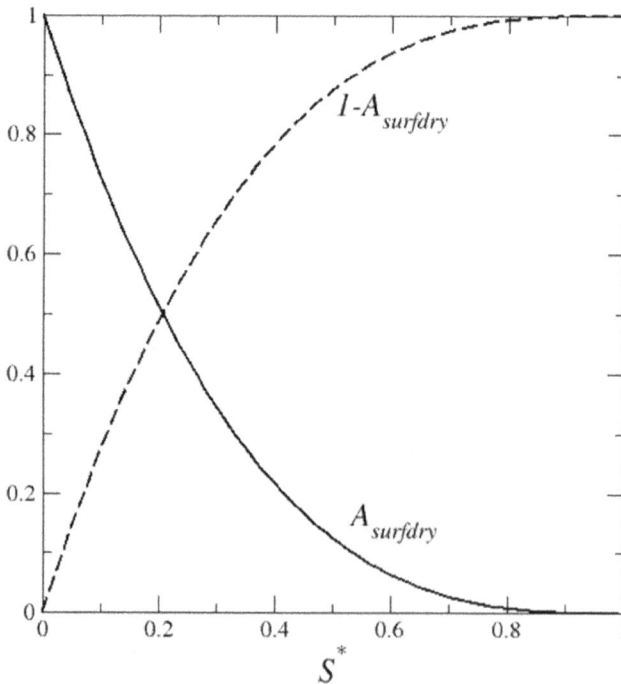

FIGURE 6.5 Variation of $A_{surfdry}$ and $A_{surfwet} = 1. - A_{surfdry}$ as a function of the reduced saturation at the surface $S^* = \left(\frac{S_{surf} - S_{irr}}{1 - S_{irr}}\right)$.

6.4.6 TBC Parameters

As regards the interfacial parameters, the situation is still more controversial than for the internal parameter η of the NLE model (Section 6.4). The interfacial parameters are S_{cr} (Eq. 6.14), r_{pm}, r_a (Eq. 6.15), $h_m(S)$, (Eq.(6.20) and S_{surf}, S_{irr}, q (Eq. 6.24), depending on the considered top boundary condition. Somewhat like the phase change parameter η, these parameters have not been defined or computed so far from well-defined closure problem in the interfacial region. The approach has been essentially empirical from comparisons with either experimental (e.g. Li et al., 2019) or PNM numerical (e.g. Talbi and Prat, 2021) simulations. In other words, the interfacial coefficients are fitted so as to obtain a reasonable agreement between continuum model simulations and drying data from experiments or from PNM simulations. Examples of such comparisons can be found in Kaviany and Mittal (1987), Chen and Pei (1989), Li et al. (2019) and Talbi and Prat (2021). This point is further discussed in the next section.

6.5 CONTINUUM MODELS VERSUS EXPERIMENTS

Naturally, the validity of the continuum models is a crucial question. Since the continuum theory of drying is incomplete, mainly because of the lack of theoretical analyses proposing a rigorous modeling of the interfacial boundary conditions and/ or allowing one to determine the interfacial condition parameters, such as the dependence of these parameters with the fluid distribution in the interfacial region, comparisons with experiments have been the commonly used approach to "validate" the continuum models.

6.5.1 LE Model Neglecting the Vapor Transport

The comparison with experiments presented in Li et al. (2019) shows that the LE model neglecting the vapor transport cannot predict the full drying process. The model underestimates the stage 2 evaporation, i.e. the period after the stage 1 is shorter than in the experiments and the final cumulative evaporation is lower than in the experiment, consistently with the fact that the vapor transport is significant in stage 2 evaporation. The fact that the model does not consider the vapor diffusion is an obvious shortcoming to simulate the stage 2 evaporation. However, it can be useful when the focus is on stage 1 evaporation. As discussed in detail in Talbi and Prat (2021), the NLE effect is confined in a thin region at the top of the porous medium in this case and the vapor transport is significant only in the interfacial region. As discussed in Section 6.5, the main difficulty is then the mass transfer parametrization at the interface (the porous medium top surface). As discussed in Talbi and Prat (2021), the interfacial resistance approach, i.e. Eq. (6.15), is attractive in this respect in relation with the thin edge effect region analyzed in Talbi and Prat (2021). However, as mention before, there is currently no theory predicting the functional form of both the interfacial and external resistances, which again both depend on the porous medium and the external flow conditions. Empirical relationships are proposed in the soil science literature (e.g. Li et al., 2019 and references therein). They should be used with caution due to their uncertain validity.

6.5.2   LE MODEL WITH VAPOR TRANSPORT

Solution to Eq. (6.10) is discussed in Pel et al. (2002). Figure 6.6 illustrates the typical evolution of the saturation profiles obtained in experiments and shown that the solution of the LE model leads to a good agreement with the data.

The top boundary condition to obtain the profiles shown in Figure 6.6 was to impose the saturation S_{top} given by the equation $\varphi\left(S_{top}\right) = RH_{inf}$. Thus, it was assumed that the relationship describing the desorption isotherm is still valid at the top surface of the porous medium during drying. However, this case is for a hygroscopic material (brick) and the comparison is performed only over the period when an internal drying front develops within the porous medium (stage 2 evaporation). This avoids the difficulty of the interfacial boundary condition when the porous medium surface is partially wet. In this respect, the favorable comparison shown in Figure 6.6 is a bit misleading since only stage 2 evaporation is considered. The simulation of the full drying process again implies to consider the problem of the top surface boundary condition.

The LE model with vapor transfer was compared with drying experiments performed with a packing of glass beads (210 μm in diameter) in Kaviany and Mittal (1987). Although drying is not quasi-isothermal in this study, this paper is representative of the current status of the LE model. A fair agreement is obtained between the model and the experiment as regards the evaporation rate evolution. However, the model involves a surface saturation coefficient in the interfacial boundary condition playing a role similar to the interfacial resistances in Eq. (6.15) or the mass transfer coefficient in Eq. (6.20). This coefficient is actually determined from the measurement of the surface temperature or the

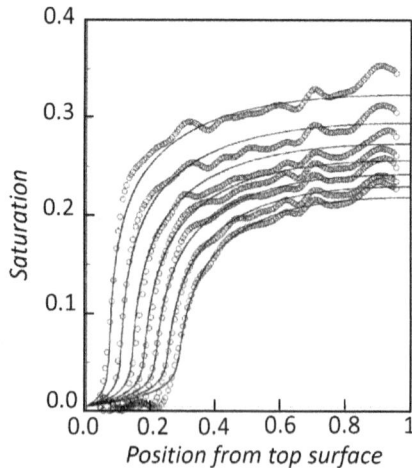

FIGURE 6.6   Typical saturation profile evolution during drying in stage 2 evaporation. The empty circles correspond to measurements. The solid lines give the saturation profiles as determined from the LE continuum model with vapor transport. (Adapted from Pel et al. (2002).)

evaporation rate. In other words, it is not predicted but adjusted. Thus, in fact, the comparison presented in Kaviany and Mittal (1987) again puts forward the lack of predictive theory regarding the surface saturation coefficient, i.e. the transfers in the interfacial region.

Predictions from the LE model with vapor transfer were compared to the experimental data from Ceaglske and Hougen (1937) for sand drying in Whitaker and Chou (1983). These authors performed the comparison trying to not use adjustable parameters. The computed drying curves for various drying conditions were not in agreement with the experimental data with a significant underprediction of the drying rates. A much better agreement was presented in a subsequent paper (Whitaker 1985) but the heat and mass transfer coefficients were fitted. The conclusion was that the "matter of the heat and mass transfer coefficients at the surface of porous medium need to be considered in greater detail". Thus, this work again points out the transfer modeling problem in the interfacial region.

The LE model with vapor transfer presented by Chen and Pei (1989) applies to both hygroscopic and capillary porous media. An original feature compared to the LE model presented in Section 6.2.2 is the consideration of bound water transport. In capillary porous media, this transport is deemed to correspond to liquid transport in "very fine capillaries". The model agrees well with the experimental data for wool, brick and corn kernels. However, the heat and mass transfer coefficients at the porous medium surface are linear functions of the moisture content at the surface involving two parameters. The latter are actually fitted parameters since they are not measured from a specific experiment or theoretically predicted. Again, this comparison cannot be considered as a true validation since the comparison is not performed in the absence of adjustable parameters.

The LE model with vapor transfer was compared to the experimental data in Li et al. (2019). It was not able to reproduce well the variation of the measured cumulative evaporation with time.

6.5.3   NLE MODEL

The NLE two-equation model was compared to the experimental data in Li et al. (2019). Like the LE model with vapor transport, it was not able to reproduce well the variation of the measured cumulative evaporation with time. The comparison was not performed in the absence of adjustable parameters. The parameters of the interfacial boundary conditions and the phase change term coefficient η (Eq. 6.13) were varied. Again, the conclusion was that future work on the interfacial boundary condition was needed.

In what follows, we present our own comparison between the NLE continuum model and the experimental data. The solution of the NLE continuum model is compared in Figure 6.7 to experimental results. The latter are drying data presented in Coussot (2000) and obtained with a packing of glass beads of diameter 17 μm contained in a cylindrical vessel. These experiments were performed with ethanol ($\gamma = 23 \times 10^{-3}$ N/m, $\mu = 1.2 \times 10^{-3}$ Pl, $\rho_l = 785$ kg/m^3, $P_{vsat.} = 7000$ Pa, $D_{vm} = 1.14 \times 10^{-5}$ m^2/s). From the data available in Coussot (2000), the thickness of the external boundary layer has been estimated so that the evaporation flux

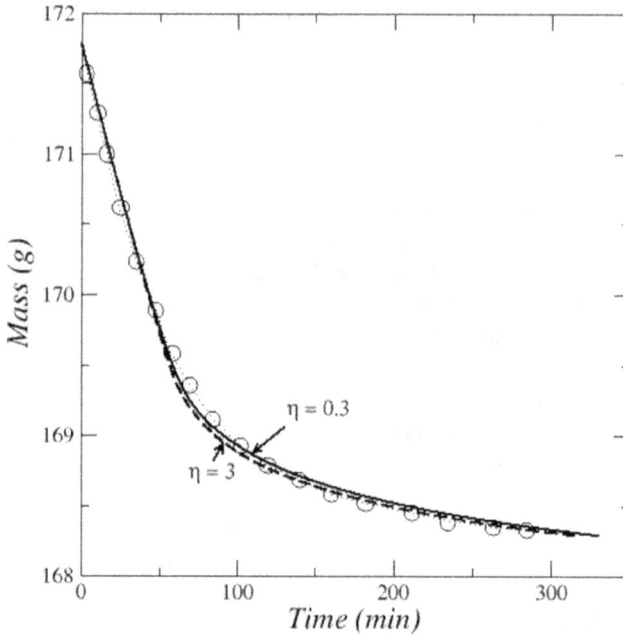

FIGURE 6.7 Mass of bead packings as a function of time. The solid line corresponds to the simulation with the NLE continuum model, whereas the circles are experimental data (Coussot, 2000).

$D_{vm} \dfrac{M_v \left(P_{vs} - P_{v\infty} \right)}{RT \; \delta}$ at the beginning of drying is the same as the experiments (note that $T = 23°C$, $P_{v\infty} = 0$). This gives $\delta = 1.11$ mm.

Figure 6.7 shows the comparison between the NLE continuum model and the experimental results from Coussot (2000). The liquid diffusivity was estimated from classical relationships for random packings of particles (Hidri et al., 2013), namely the Kozeny–Carman relationship, $k = \dfrac{\varepsilon^3 d^2}{180 \left(1 - \varepsilon \right)^2}$, $k_r = \left(\dfrac{S - S_C}{1 - S_C} \right)^3$,

$\dfrac{S - S_C}{1 - S_C} = \dfrac{1}{\left[1 + \left(\dfrac{P_c}{P_{cref}} \right)^n \right]^m}$ with n = 8 and $m = 1 - \dfrac{1}{n}$ $P_{cref} = \dfrac{6(1-\varepsilon)\gamma}{d}$ where γ is the surface tension and P_{cref} is the air entry pressure. The NLE continuum model simulation results were obtained for $S_c = 0.00001$, a much lower value than the value classically considered in drainage ($S_c = 0.1$).

As illustrated in Figure 6.7, the NLE model is able to reproduce the experimental data reported in Coussot (2000) on drying of a glass bead packing. However, in addition to S_c, parameter η (Eq. 6.15) was varied so as to obtain the good agreement. It was found that η in the range [0.3–3] led to the reasonable agreement shown in Figure 6.7. Thus, again, the comparison was performed considering adjustable parameters. Further tests are needed to evaluate whether the NLE continuum model is truly superior to the LE model.

6.6 A MORE ADVANCED CONTINUUM MODEL

As discussed in Prat (2002) and Talbi and Prat (2022), a key feature of the drying process is the gradual fragmentation of the liquid phase in liquid clusters during the drying process. This is illustrated in Figure 6.8 from PNM simulations (Le Bray and Prat, 1999).

As can be seen, the number of liquid clusters increases in a first period corresponding to stage 1 evaporation and the beginning of stage 2 evaporation. Then, the number of clusters decreases as an internal drying front forms and gradually recedes within the porous medium. Also, the cluster distribution is characterized in stage 1 evaporation by the presence of a percolating liquid cluster and numerous non-percolating smaller liquid clusters. By definition, the percolating liquid cluster, also referred to as the main cluster (Le Bray and Prat, 1999; Yiotis et al., 2006), is connected to both the porous medium bottom and top surfaces. As discussed in Talbi and Prat (2022), see also Ahmad et al. (2021), the fact that the liquid phase is fragmented has a great impact of the distribution of solute of particles within the medium during drying. It is thus interesting to explore whether this key aspect can be explicitly considered within the framework of continuum models. This can also be related to the two-phase flow theory by Hilfer and colleagues (Doster et al., 2010) in which an explicit distinction is made between the percolating and non-percolating phases. The main novelty with the three-equation continuum model is to explicitly consider that the liquid phase can be split into the percolating liquid

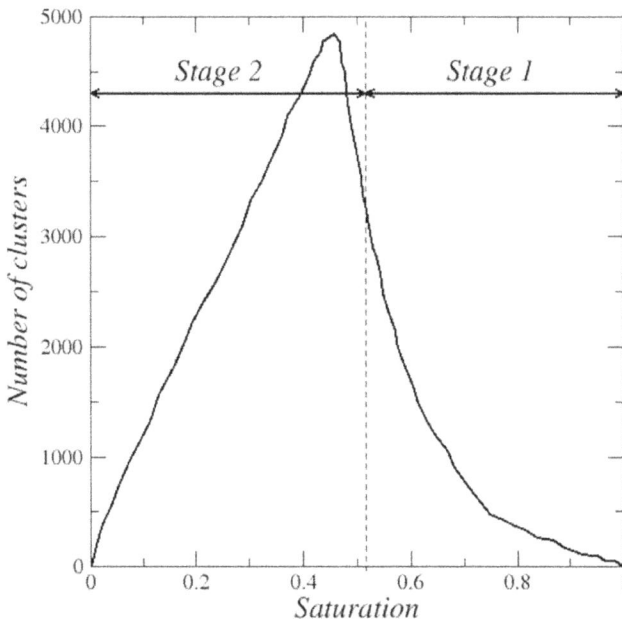

FIGURE 6.8   Number of liquid clusters as a function of the overall saturation during drying. (Adapted from Le Bray and Prat (1999).)

phase and the non-percolating liquid phase. In the three-equation model, the liquid saturation is thus expressed as

$$S = S_1 + S_2 \tag{6.25}$$

where subscript 1 is for the percolating liquid phase, i.e. the main cluster, and subscript 2 for the non-percolating liquid phase, i.e. the isolated clusters. Subscript 3 refers to the vapor.

Assuming a homogeneous porous medium, the mass balance equation for the percolating liquid phase is expressed as

$$\varepsilon \rho_l \frac{\partial S_1}{\partial t} + \nabla.\left(\rho_l D_{l1}(S_1)\nabla S_1\right) = -\dot{m}_{12} - \dot{m}_{13} \tag{6.26}$$

in which $D_{l1}(S_1) = -\dfrac{kk_{r1}}{\mu}\dfrac{dP_{c1}}{ds_1}$, where k is the medium permeability; k_{r1} is the percolating phase relative permeability; μ is the liquid viscosity; $P_{cl}(S_l)$ is the local pressure difference between the gas phase and the percolating liquid phase, i.e. the capillary pressure curve considering only the main cluster.

The non-percolating liquid phase is assumed to be immobile. The non-percolating liquid phase mass conservation equation thus reduces to

$$\varepsilon \rho_l \frac{\partial S_2}{\partial t} = -\dot{m}_{21} - \dot{m}_{23} \tag{6.27}$$

In the above equations, ε is the porous medium porosity, ρ_l is the liquid density, $\dot{m}_{12}$ is the mass transfer rate between phase 1 and phase 2, $\dot{m}_{13}$ is the evaporation rate of phase 1 per unit volume of porous medium, $\dot{m}_{21}$ is the mass transfer rate between phase 2 and phase 1, $\dot{m}_{23}$ is the evaporation rate of phase 2 per unit volume of porous medium. Since a new isolated cluster actually forms as a result of the fragmentation of the main cluster,

$$\dot{m}_{12} = -\dot{m}_{21} \tag{6.28}$$

It can be noted that in isothermal drying, an isolated cluster cannot reconnect to the main cluster, hence $\dot{m}_{12} > 0$.

The gas phase forms a single cluster in the drying process. The mass conservation of the vapor is expressed as for the NLE two-equation continuum model, i.e. Eq. (6.12):

$$\nabla.\left(\varepsilon(1-S)D_{eff}\frac{M_v}{RT}\nabla P_v\right) + \dot{m} = 0 \tag{6.29}$$

where

$$\dot{m} = \dot{m}_{13} + \dot{m}_{23} \tag{6.30}$$

The phase change rate is expressed for the NLE two-equation continuum model as

$$\dot{m} = a_{lg}\eta\frac{M_v}{RT}(P_{vs} - P_v)$$ (6.31)

with

$$\dot{m}_{13} = a_{l1g}\eta\frac{M_v}{RT}(P_{vs} - P_v)$$ (6.32)

$$\dot{m}_{23} = a_{l2g}\eta\frac{M_v}{RT}(P_{vs} - P_v)$$ (6.33)

where a_{l1g} (a_{l2g} respectively) is the specific interfacial area between phase 1 (phase 2 respectively) and the gas phase. It can be noted that $a_{lg} = a_{l1g} + a_{l2g}$.

Following Doster et al. (2010), the mass transfer rate $\dot{m}_{12}$ between the percolating and non-percolating liquid phases is expressed as

$$-\dot{m}_{12} = \chi\varepsilon\rho_l\left(\frac{S_2 - S_{irr}}{S_{irr} - S}\right)\frac{\partial S}{\partial t}$$ (6.34)

where χ is a numerical factor and S_{irr} is the irreducible saturation.

The key new relationship in this model is Eq. (6.34), which takes into account the fragmentation process, i.e. the formation of isolated clusters from the main (percolating) cluster, within the continuum approach framework. As reported in Talbi and Prat (2022), this model has been tested against pore network simulations considering a special case allowing to focus on the impact of Eq. (6.34). This special case corresponds to stage 1 evaporation when capillary effects are dominant. The NLE effect is then confined in the interfacial region, i.e. the region of the porous medium in contact with the surrounding air, and the saturation S_1 and S_2 are spatially uniform, which greatly simplifies the computation of the saturation variations.

Figure 6.9 shows the typical evolution of S, S_1 and S_2 in this regime during stage 1 evaporation as computed from the above equations in the considered drying regime. As reported in Talbi and Prat (2022), the comparison between the three-equation continuum model and the PNM simulations is quite good, provided that the finite effects due to the network small size are adequately considered. This validates the expression of the percolating–non-percolating liquid phase exchange term adapted from Doster et al. (2010), i.e. Eq. (6.34). However, it should be noted that the numerical factor χ was adjusted to perform the comparison with the PNM simulation results. In other words, here also a specific closure problem enabling one to determine χ for a given microstructure is lacking for considering the three-equation model as fully predictive. The same holds for the evaluation of η as discussed previously. By contrast, the computation of $D_{l1}(S_1)$ is presumably less an issue and could be performed for example by PNM simulations over a representative elementary volume adapting the algorithms used to determine the classical capillary curve and the relative permeabilities (Bunt et al., 2002). This also holds for a_{l1g} and a_{l2g}, using similar approaches as for a_{lg} (Joekar-Niasar et al., 2008).

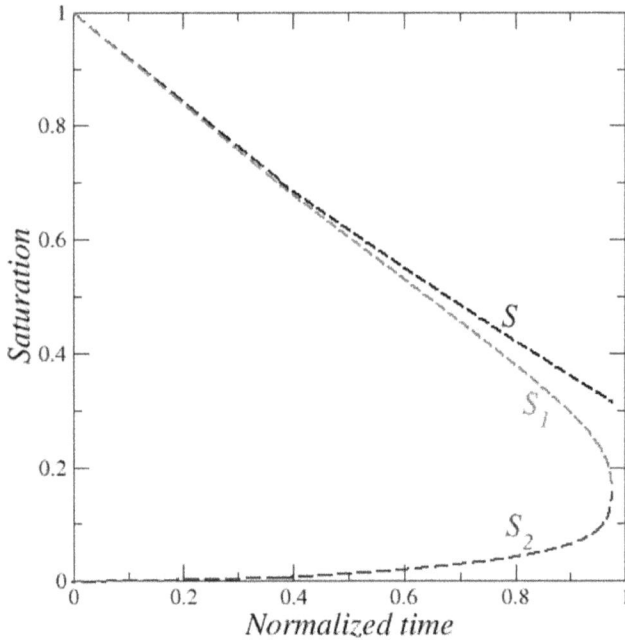

FIGURE 6.9   Typical variations of S, S_1 and S_2, as a function of time during stage 1 evaporation. (Adapted from Talbi and Prat (2021).)

As for the other continuum models, a major issue lies in the interfacial boundary conditions. As for the NLE two-equation model, one should take into consideration that the evaporation at the surface can be separated into a contribution from the wet surface pores and one from the dry surface pores with the additional difficulty that a fraction of the wet surface pores belongs to the percolating liquid phase and the complementary fraction to the non-percolating liquid phase. These difficulties were circumvented in Talbi and Prat (2022) via the consideration of only stage 1 evaporation and the so-called capillary regime. For a more general use of the three-equation model, the question of the interfacial boundary condition remains an open topic. It can be also noticed that the three-equation model has not yet been compared to the experimental data.

6.7   DISCUSSION

The brief literature review presented in Section 6.6 shows that the status of the drying continuum modeling has not changed very much since the eighties. The continuum models are able to reproduce important features of the drying process such as the saturation profiles, the impact of the drying rate on the saturation profiles, the pore size impact (e.g. Coussot, 2000) and the main drying periods (or stages). However, the drying continuum theory is not complete since it is not free of adjustable parameters. For this reason, this is not a truly predictive theory. Regarding the internal transport, two parameters have been used as adjustable parameters. The first one is the liquid relative permeability in the range of the very low saturations (e.g. Whitaker, 1985; Hadley, 1985). The positive effect of increasing the liquid relative permeability in an extended range of low saturations could be related to the impact of liquid films. Experiments (Laurindo and Prat, 1998; Yiotis et al., 2012) and PNM simulations (Yiotis et al., 2004; Prat, 2007) have shown that capillary film flows in corners could have a quite significant impact of drying. Unlike the adsorbed films in the hygroscopic porous media, referred to as bound water films(i.e. Chen and Pei, 1989), capillary films are not explicitly considered in the drying continuum models. The results of Whitaker and Hadley suggest that the effect of capillary films might be taken into account via the liquid diffusivity, which is the key parameter in the modeling of the liquid flow in the continuum models. However, in the present state of the art, this is just an empirical way for including a film effect. In fact, the modeling of the film effects within the continuum approach framework would deserve to be considered in more depth. In this respect, the modeling of the film effects in the PNM (Yiotis et al., 2004; Prat, 2007) could serve as guidelines. The second adjustable internal parameter is the phase change parameter χ (Eq. 6.13). Although experimental determinations have been performed (e.g. Lozano et al., 2008), a more general study of the dependence of this parameter with the porous medium microstructure geometrical properties is needed.

Nevertheless, the most difficult topic remains the parametrization of the transfers in the interfacial regions of the porous medium. As discussed in this chapter, various formulations have been used in the literature. All involve adjustable parameters and lack fundamental basis. Actually, the coupling at the interface cannot be simply expressed as a "simple" conjugate transfer problem between the continuum formulation of the transport phenomena in the porous medium and the external transfer formulation, i.e. the continuity of the macroscopic vapor partial pressure (and temperature) and macroscopic fluxes at a surface of zero thickness (e.g. Defraeye et al., 2012). PNM simulations (e.g. Talbi and Prat, 2021) make clear that the transfers at the surface are affected by both the fluid distribution in the interfacial region, which is different from deeper in the porous medium, and the mass transfer rate. Furthermore, the PNM simulations indicate a significant NLE in the interfacial region. As a result, the development of a fully predictive theory of the transfers in the interfacial region remains a challenge.

Another issue requiring clarification is the impact of liquid bridges (e.g. Vorhauer et al., 2015) and water in pendular states at very low saturations in capillary porous media, i.e. the transfer mechanisms at very low saturation. These effects can be considered in PNM (Vorhauer et al., 2015; Kharaghani et al., 2021). However, they were not taken into account in the comparisons performed so far between PNM simulations and continuum models (e.g. Attari-Moghaddam et al., 2017b; Ahmad et al., 2020).

Throughout this chapter, the gas phase total pressure was considered uniform and constant over the gas phase. This assumption is not valid in the case of a porous medium submitted to high heat fluxes (Constant et al., 1996) but is commonly made for modeling the slow drying process considered in this chapter. However, this assumption has been questioned even for slow drying conditions by Mainguy et al.

(2001) in their analysis of the drying of concrete. This topic is controversial; since a few years later, Thiery et al. (2007) reached the conclusion that the gas total pressure variation could not be inferred from the data presented in Mainguy et al. (2001). Nevertheless, this issue is perhaps not completely clarified and the fact that the total pressure could play a role in the drying process of, at least, hygroscopic materials even under slow drying conditions can be kept in mind. It seems, however, that the adequate consideration of corner film flows and bound water flows within the framework of the continuum approach is of higher priority as regards the modeling of slow drying.

6.8 SUMMARY

Continuum models are by far the most adequate models to investigate by means of numerical simulations of the drying process in capillary porous media in most applications. Very often, their computational performances make them the only option in practice compared to pore scale models such as pore network models (Prat, 2002) or direct simulations (e.g. Panda et al., 2020). They reproduce quite well many important features of the drying process. However, the continuum approach to drying remains incomplete, mainly due to the lack of fully predictive and rigorous theory of the transfers in the interfacial region with the surrounding air. Although dedicated studies (e.g. Shahraeeni et al., 2012b; Mosthaf et al., 2014; Talbi and Prat, 2021, among others) have brought numerous insights on the transfers in the interfacial region, the modeling of the interfacial transfers within the continuum approach framework remains a widely open question. This is the most important missing piece in the continuum theory of drying. In this respect, the drying modeling problem can be related to the general problem of the interfacial conditions between a free fluid and a porous medium (e.g. Beavers and Joseph, 1967; Ochoa-Tapia and Whitaker, 1997; Valdés-Parada et al., 2006; Chandesris and Jamet, 2009). However, the latter references only address single-phase transfer in the pores. The drying problem is presumably significantly more difficult owing to the evolution of the fluid distribution in the interfacial region during drying. Nevertheless, the aforementioned works, and references therein, could be a good basis to develop more satisfactory interfacial relationships. In spite of this limitation, drying continuum models remain as quite valuable tools to study realistic drying processes in porous media.

As presented in this chapter, there is not a single drying continuum model but several variants depending on the consideration or not of the internal vapor transfer, the consideration or not of the local equilibrium assumption and the formulation of the interfacial boundary conditions. This actually offers some flexibility in the choice of the model depending on the considered objective. For instance, as exemplified in Talbi and Prat (2021, 2022), the simple LE model in the absence of vapor transport can be sufficient if the focus is on stage 1 evaporation.

ACKNOWLEDGMENT

Financial support from joint project "Drycap" funded by GIP ANR (project16-CE92-0030-01) and DFG (project TS28/10–1) is gratefully acknowledged.

REFERENCES

Adamson A.W. 1990. *Physical Chemistry of Surfaces*, 5th ed., Wiley, New York.

Ahmad, F., Rahimi, Tsotsas, E., Prat, M., Kharaghani, A. 2021. From micro-scale to macro-scale modeling of solute transport in drying capillary porous media, *Int. J. Heat Mass Transf.* 165: 120722.

Ahmad, F., Talbi, M., Prat, M., Tsotsas, E., Kharaghani, A. 2020. Non-local equilibrium continuum modeling of partially saturated drying porous media: Comparison with pore network simulations. *Chem. Eng. Sci.*, 228: 115957.

Attari-Moghaddam, A., Kharaghani, A., Tsotsas, E., Prat, M. 2017a. Kinematics in a slowly drying porous medium: Reconciliation of pore network simulations and continuum modeling, *Phys. Fluids* 29(2): 022102.

Attari-Moghaddam, A., Prat, M., Tsotsas, E., Kharaghani, A. 2017b. Evaporation in capillary porous media at the perfect piston-like invasion limit: Evidence of non-local equilibrium effects, *Water Resour. Res.* 53(12): 10433–10449.

Auriault, J.L., Boutin, C., Geindreau C., 2010. *Homogenization of Coupled Phenomena in Heterogenous Media*, John Wiley & Sons, Hoboken, USA.

Beavers, G.S., Joseph, D.D. 1967. Boundary conditions at a naturally permeable wall. *J. Fluid Mech.* 30(1): 197–207.

Benet, J.C., Jouanna, P. 1982. Phenomenological relation of phase change of water in a porous medium: Experimental verification and measurement of the phenomenological coefficient. *Int. J. Heat Mass Transf.* 25: 1747–1754.

Blunt, M.J., Jackson, M.D., Piri, M., Valvatne, P.H., 2002. Detailed physics, predictive capabilities and macroscopic consequences for pore-network models of multiphase flow. *Adv. Water Resour.* 25(8–12), 1069–1089

Bouddour, A., Auriault, J.L., Mhamdi-Alaoui, M., Bloch, J.F. 1998, Heat and mass transfer in wet porous media in presence of evaporation—condensation. *Int. J. Heat Mass Transf.* 41(15): 2263–2277.

Ceaglske, N.H., Hougen, O.A. 1937. Drying granular solids. *Ind. Eng. Chem.* 29: 805–813.

Chandesris, M., Jamet, D. 2009. Jump conditions and surface-excess quantities at a fluid/porous interface: A multi-scale approach. *Transport Porous Med.* 78(3): 419–438.

Chen, P., Pei, D.C.T. 1989. A mathematical model of drying process. *Int. J. Heat Mass Transf.* 32(2): 297–310.

Constant, T., Moyne, C., Perré P. 1996. Drying with internal heat generation: Theoretical aspects and application to microwave heating. *AIChE J.*, 42(2): 359–368.

Coussot, P. 2000, Scaling approach of the convective drying of a porous medium. *Eur. Phys. J. B* 15: 557–566.

Culligan, K.A., Wildenschild, D., Christensen, B.S.B., Gray, W.G., Rivers M.L., Tompson, A.F.B, 2004. Interfacial area measurements for unsaturated flow through a porous medium. *Water Resour. Res.* 40, W12413.

Defraeye, T., Blocken, B., Derome, D., Nicolai, B., Carmeliet, J. 2012. Convective heat and mass transfer modeling at air-porous material interfaces: Overview of existing methods and relevance, *Chem. Eng. Sci.* 74: 49–58.

Doster, F., Zegeling, P.A., Hilfer, R. 2010. Numerical solutions of a generalized theory for macroscopic capillarity. *Phys. Rev. E* 81: 036307.

Dullien, F.A.L. 1992. *Porous Media: Fluid Transport and Pore Structure*. Academic Press, San Diego, USA.

Guibert, R., Horgue, P., Debenest, G., Quintard M. 2016. A comparison of various methods for the numerical evaluation of porous media permeability tensors from pore-scale geometry. *Math. Geosci.* 48: 329–347.

Hadley, G.R. 1985. Numerical modeling of the drying of porous materials. In: Toei R., Mujumdar A.S. (eds) *Drying '85*. Springer, Berlin, Heidelberg.

Hidri, F., Sghaier, N., Eloukabi, H., Prat, M., Ben Nasrallah, S. 2013. Porous medium coffee ring effect and other factors affecting the first crystallization time of sodium chloride at the surface of a drying porous medium. *Phys. Fluids* 25: 127101.

Joekar-Niasar, V., Hassanizadeh, S.M., Leijnse A. 2008. Insights into the relationships among capillary pressure, saturation, interfacial area and relative permeability using pore-network modeling. *Transp. Porous Med.* 74: 201–219.

Kaviany, M. 1991, *Principles of Heat Transfer in Porous Media*. Springer, New York.

Kaviany, M., Mittal, M. 1987. Funicular state in drying of a porous slab. *Int. J. Heat Mass Transf.* 30(7): 1407–1418.

Kays, W. M., Crawford, M.E. 1980. *Convective Heat and Mass Transfer*. McGraw-Hill, New York.

Kharaghani, A., Mahmood, H.T., Wang, Y., Tsotsas, E. 2021, Three-dimensional visualization and modeling of capillary liquid rings observed during drying of dense particle packings, *Int. J. Heat Mass Transf.* 177: 121505.

Krischer, O. 1938. Fundamental law of moisture movement in drying by capillary flow and vapor diffusion. *V.D.I Z.* 82: 373–378.

Laurindo, J.B., Prat, M. 1998. Numerical and experimental network study of evaporation in capillary porous media. Drying rates. *Chem. Eng. Sci.* 53(12): 2257–2269.

Le Bray, Y., Prat, M. 1999. Three-dimensional pore network simulation of drying in capillary porous media. *Int. J. Heat Mass Transf.* 42: 4207–4224.

Lehmann, P., Assouline, S., Or Dani, 2008. Characteristic lengths affecting evaporative drying of porous media. *Phys. Rev. E* 77, 056309

Li, Z., Vanderborght, J., Smits, K.M. 2019. Evaluation of model concepts to describe water transport in shallow subsurface soil and across the soil–air interface. *Trans. porous Med.* 128(3): 945–976.

Lozano, A.L., Cherblanc, F., Cousin, B., Bénet, J.C., 2008. Experimental study and modeling of the water phase change kinetics in soils. *Eur. J. Soil Sci.* 59: 939–949.

Luikov, A.V. 1966. *Heat and Mass Transfer in Capillary Porous Bodies*. Oxford, Pergamon.

Maalal, O., Prat, M., Lasseux, D. 2021. Pore network model of drying with Kelvin effect, *Phys. Fluids* 33(2): 027103.

Mainguy, M., Coussy, O., Baroghel-Bouny, V. 2001. Role of air pressure in drying of weakly permeable materials. *J. Eng. Mech.* 127: 582–592.

Masmoudi, W., Prat, M., 1991. Heat and mass transfer between a porous medium and a parallel external flow, application to drying of capillary porous materials, *Int. J. Heat Mass Transf.* 34(8): 1975–1989.

Millington, R.J., Quirk, J.P., 1961. Permeability of porous solids, *Trans. Faraday Soc.* 57: 1200–1207.

Mosthaf, K., Helmig, R., Or, D. 2014. Modeling and analysis of evaporation processes from porous media on the REV scale. *Water Resour. Res.* 50: 1059–1079.

Ochoa-Tapia, J.A., Whitaker, S. 1997. Heat transfer at the boundary between a porous medium and a homogeneous fluid. *Int. J. Heat Mass Transf.* 40(11): 2691–2707.

Ouedraogo, F., Cherblanc, F., Naon, B., Bénet, J.C. 2013, Water transfer in soil at low water content. Is the local equilibrium assumption still appropriate? *J. Hydrol.* 492: 117–127.

Panda, D., Paliwal, S., Sourya, D.P., Kharaghani, A., Tsotsas, E., Surasani, V.K. 2020. Influence of thermal gradients on the invasion patterns during drying of porous media: A lattice Boltzmann method. *Phys. Fluids* 32: 122116.

Pel, L., Ketelaars, A.A.J., Adan, O.C.G., Van Well, A.A. 1993. Determination of moisture diffusivity in porous media using scanning neutron radiography, *Int. J. Heat Mass Transf.* 36(5): 1261–1267.

Pel, L., Landman, K.A., Kaasschieter, E.F. 2002. Analytic solution for the non-linear drying problem, *Int. J. Heat Mass Transf.* 45(15): 3173–3180.

Perré, P., Turner, I.W. 1999. A 3-D version of TransPore: A comprehensive heat and mass transfer computational model for simulating the drying of porous media. *Int. J. Heat Mass Transf.* 42(24): 4501–4521.

Philip, J.R., de Vries D.A. 1957. Moisture movement in porous materials under temperature gradients. *Trans. Am. Geophys. Union* 38: 222–232.

Plourde, F., Prat, M. 2003. Pore network simulations of drying of capillary media. Influence of thermal gradients. *Int. J. Heat Mass Transf.* 46: 1293–1307.

Prat, M. 2002. Recent advances in pore-scale models for drying of porous media. *Chem. Eng. J.* 86: 153–164.

Prat, M. 2007. On the influence of pore shape, contact angle and film flows on drying of capillary porous media. *Int. J. Heat Mass Transf.* 50: 1455–1468.

Schlichting, H., Gersten, K. 2016. *Boundary-Layer Theory.* 9th Edition, Springer, Berlin and Heidelberg.

Shahraeeni, E., Lehmann, P., Or, D. 2012b. Coupling of evaporative fluxes from drying porous surfaces with air boundary layer: Characteristics of evaporation from discrete pores. *Water Res. Res.* 48: W09525.

Shahraeeni, E., Or, D. 2012a. Pore scale mechanisms for enhanced vapor transport through partially saturated porous media. *Water Resour. Res.* 48, W05511.

Sherwood, T.K. 1929. The drying of solids. *Ind. Eng. Chem.* 21: 12–16.

Shokri, N., Lehmann, P., Or, D. 2009. Critical evaluation of enhancement factors for vapor transport through unsaturated porous media. *Water Resour. Res.* 45: W10433.

Talbi, M., Prat, M. 2021. Coupling between internal and external mass transfer during stage-1 evaporation in capillary porous media: Interfacial resistance approach. *Phys. Rev. E* 104, 055102.

Talbi, M., Prat, M. 2022. Percolating and non-percolating liquid phase continuum model of drying in capillary porous media with application to solute transport in the very low Péclet number limit. *Phys. Rev. Fluids* 7, 014306.

Thiery, M., Baroghel-Bouny, V., Bourneton, N., Villain, G., Stéfani C. 2007. Modélisation du séchage des bétons, analyse des différents modes de transfert hydrique. *Revue européenne de Génie Civil* 11: 541–577.

Valdés-Parada, F.J., Goyeau, B., Ochoa-Tapia, J.A. 2006. Diffusive mass transfer between a microporous medium and an homogeneous fluid: Jump boundary conditions, *Chem. Eng. Sci.* 61(5): 1692–1704.

Van Brakel, J. 1980. Mass transfer in convective drying. *Adv. Dry.* 1: 217–267.

Vorhauer, N., Wang, Y., Karaghani, A., Tsotsas, E., Prat, M. 2015. Drying with formation of capillary rings in a model porous medium. *Trans. Porous Med.* 110: 197–223.

Vu, H.T., Tsotsas E. 2018. Mass and heat transport models for analysis of the drying process in porous media: A review and numerical implementation, *Int. J. Chem. Eng.*, Vol. 2018: ID 9456418.

Whitaker, S., James P. Hartnett., Thomas F. Irvine. 1977. Simultaneous heat, mass and momentum transfer in porous media. A theory of drying. In: *Advances in Heat Transfer*, Vol. 13, pp. 119–203, Academic Press, New-York.

Whitaker, S. 1985. Moisture transport mechanisms during the drying of granular porous media. In: Toei R., Mujumdar A.S. (eds) *Drying '85,* p. 21–32. Springer, Berlin, Heidelberg.

Whitaker, S. 1986. Flow in porous media II: The governing equations for immiscible, two-phase flow, *Trans. Porous Med.* 1(2): 105–125.

Whitaker, S., 2013. *The Method of Volume Averaging*, Springer Science & Business Media, Dordrecht.

Whitaker, S., Chou, W.T.H. 1983. Drying granular porous media-theory and experiment. *Dry. Technol.* 1(1): 3–33.

Wu, Q., Rougelot, T., Burlion N., Bourbon, X. 2015. Representative volume element estimation for desorption isotherm ofconcrete with sliced samples. *Cem. Concr. Res.* 76: 1–9.

Yiotis, A.G., Boudouvis, A.G., Stubos, A.K., Tsimpanogiannis, I.N., Yortsos, Y.C. 2004. Effect of liquid films on the drying of porous media. *AIChE J.* 50: 2721–2737.

Yiotis, A.G., Salin, D., Tajerand, E.S., Yortsos, Y.C. 2012. Drying in porous media with gravity-stabilized fronts: Experimental results. *Phys. Rev. E* 86: 026310.

Yiotis, A.G., Tsimpanogiannis, I.N., Stubos, A.K., Yortsos, Y.C. 2006. Pore-network study of the characteristic periods in the drying of porous materials. *J. Colloid Interface Sci.* 297(2): 738–748.

7 A Continuum Approach to the Drying of Small Pore Networks

Xiang Lu, Evangelos Tsotsas, and Abdolreza Kharaghani
Otto-von-Guericke-Universitat Magdeburg

CONTENTS

DOI: 10.1201/9781003011811-7

7.1 INTRODUCTION

The most commonly used approach for describing the characteristics of a drying porous medium approximates the medium by a continuum within which the dynamics of heat and mass transfer processes is described using associated macroscopic equations. This modeling approach dates back to the work of Philip and Vries (1957) in which they proposed a diffusion-based equation to describe the moisture transport in porous media during evaporation. Later, advanced continuum models (CMs) are developed using the method of volume averaging. Three phases (vapor, liquid, and solid) were coupled in the transport equations, and these equations are volume averaged to provide a rational route to a set of equations describing the transport of heat and mass in a porous medium (Whitaker, 1977). Those general equations can be applied to describe, for example, the drying process in packed beds that is in meters range or the movement of water inside the gas diffusion layer in a fuel cell that is in the micrometers.

Like any other CM relevant to fluid transport in porous media, the CMs for drying require non-linear parameters which describe the fluid transport phenomena (permeability, moisture transport coefficient, vapor diffusivity, and mass transfer coefficient) and the local equilibrium (sorption isotherm). The physical phenomena that occur at the pore scale are essentially concealed in these parameters. The mass transfer coefficient introduced to describe the mass transfer at the interface between the medium surface and the bulk air serves as an example. The coefficient is often determined empirically (Bird et al., 2007), leading to a good prediction of drying. In this view, the mass transfer coefficient is a fitting parameter which yields simulation results being in sufficient agreement with datasets obtained from experiments. However, the actual scenario as regards the surface phenomena is more complex than generally assumed. In fact, the behavior of the surface reflects the complex interplay between the internal fluid flow and the adjacent gas-side boundary layer. Therefore, one can impose flux boundary conditions at the interface between the medium surface and the adjacent gas-side boundary layer to calculate the mass transfer coefficient. In this case, there is no need for any empirical correlation or fitting.

Another issue in the CMs for drying lies in the determination of the surface vapor pressure as a function of saturation. In the context of drying hygroscopic materials, desorption isotherms are often used to link the saturation and vapor partial pressure. However, the application of this concept for drying capillary porous materials, which are usually considered as non-hygroscopic porous materials composed of pores with sizes in the order of 100 µm, can be questioned (Geoffroy and Prat, 2014; Maalal et al., 2021). Recently, this aspect of drying has vastly been investigated by means of results obtained from extensive pore network simulations (Attari Moghaddam, 2017; Lu, 2021). Some results with regard to this aspect will be presented in this chapter.

From the modeling perspective, the vapor pressure at the surface of a capillary porous medium is often simply assumed as the saturation vapor pressure when the surface contains liquid (i.e. during the first drying period). When the drying front (DF) recedes during the second drying period, the surface vapor pressure is determined by the desorption isotherms (Pel et al., 1993; Pel and Landman, 2004). This assumption leads to a good prediction for the first drying period. On the other hand, it rarely predicts accurately the transition from the first drying period to the second drying period. This might be due to the fact that the previous CMs neglect the development of a dry zone (DZ) evolving from the DF to the surface of the porous medium. This DZ is filled with only vapor, the transport of which is determined by the difference in its partial pressure. The vapor transport in this zone can be described by a diffusion-based equation. In this way, a drying porous medium is divided into three transport zones: The wet zone (saturated or two-phase flow zone), the DZ (formed after the DF recedes), and the gas-side boundary layer, leading to three separate governing equations. To integrate these three zones in one CM, flux boundary conditions are imposed at two interfaces: The interface between the gas-side boundary layer and the medium-side dry region, as well as the interface between the dry region and the partially saturated part of the wet region which evolves freely during drying. It should be noted that these three zones can also be seen in the pore network simulations of drying. The required parameters in this new CM can alternatively be determined from the pore network simulations.

The pertinent question that now arises is how the macroscopic parameters of the CMs for drying can be estimated. In general, the determination of these parameters is key to the CM, but not easy. Pel et al. (1993, 1996) carried out drying experiments with brick and kaolin clay and determined the saturation profiles based on scanning neutron radiography. Then, they solved an inverse problem to obtain the moisture transport coefficient (D). Based on the variation of moisture diffusivity with moisture content, three regimes were distinguished. At high saturation, moisture transport is dominated by liquid transport. With decreasing moisture, the large pores are drained by the gas phase and they do not contribute any more to liquid transport, so that moisture diffusivity decreases. When the liquid within the sample no longer forms a continuous phase, moisture transport is governed by vapor pressure. Gómez et al. (2007) calculated permeability for the liquid phase and for the vapor phase by recording the capillary pressure profile with global saturation along the drying process. The moisture diffusivity is then calculated based on the values of these permeabilities. In addition to these experimental methods, the macroscopic parameters shall be computed based on images of a drying sample acquired using visualization techniques with higher spatial resolutions, such as X-ray microtomography (Wang et al., 2012; Kharaghani et al., 2021), confocal laser scanning microscopy (Vorhauer et al., 2015), or scanning electron microscopy (Maier et al., 2021).

Pore network models (PNMs) offer an alternative to tedious experiments. They are developed to describe transport phenomena in partially saturated porous media at the pore scale. A PNM approximates the void space within a porous medium as a network of pores and nodes. Simple physical laws applied to each pore result in large systems of algebraic equations with degrees of freedom at each node of the pore network. In early PNMs of drying by Prat (1993), the drying process is treated as an invasion percolation phenomenon (Wilkinson and Willemsen, 1983) driven by evaporation. Later on, more versatile PNMs were developed by several research groups (Yiotis et al., 2001; Metzger and Tsotsas, 2008; Wu et al., 2020, Kharaghani et al., 2021). Yiotis et al. (2001, 2015) treated the void space in the porous medium as spherical pore bodies, leaving cylindrical throats as conductors and capillary barriers. This approach is suitable for materials with large pore volumes, such as granular materials. Prat (1998), Le Bray and Prat (1999), and Metzger et al. (2007) approximated the void space using hydraulic pore throats and numerical pore nodes – this model is referred to as the throat-node model (TNM) in this chapter. Although this approach extended the application range of PNMs, it has the drawback of void space being overestimated. In the TNM, the void space is calculated by summing all throat volumes, which are considered as cylinders with fixed length. Based on the assumption that the throats are connected with numerical nodes, there is a region of overlap at each node. If one adds all throat volumes in TNM to yield the void space of a porous medium while ignoring the overlap region, then the void space is overestimated. Recently, Lu et al. (2020) used the throat-pore model (TPM) in which both cylindrical throats and spherical pores contain liquid and contribute to moisture transport. Despite such progress made in the development of PNMs, they are prone to limitations as regards representing the actual geometry of the void space and also describing the underlying transport mechanisms. One way to overcome these limitations is to opt for the Lattice Boltzmann method (LBM) models of drying (Zachariah et al., 2019; Panda et al., 2020). However, the LBM models often require high computational cost.

As regards the macroscopic parameters, Attari Moghaddam et al. (2017) evaluated the moisture transport coefficients by presenting three non-linear curves over the whole drying process. In each curve, two parts were distinguished: Values scattered when the local saturation was larger than 0.68 but became clear for local saturations smaller than 0.68. Possible reasons behind the scatter of data have not been discussed in that work. In contrast, Lu et al. (2020) treated the dataset from PNM with finer intervals of network saturation and divided the entire drying process into seven curves. They indicated that the moisture transport coefficient increases with reduced local saturation because liquid viscosity prevails at the bottom which contains high saturation. Additionally, other macroscopic parameters, i.e. the vapor diffusion coefficient at the surface and in the gas-side boundary layer, were assessed by Lu et al. (2021a, 2021b). They obtained the parameters of the CM from pore network simulations conducted with a network size of 25×25×50 pores and for the gas-side boundary layer of size 25×25×10.

The purpose of this chapter is to create pore-scale resolved data for the drying porous medium by pore network modeling and then derive macroscopic parameters from such data which can be used in the frame of CM for quick computations at the macroscale. The treatment of macroscopic parameters from the small network here differs from the classical upscaling approach which is to derive the transport coefficients on a coarse scale in which the capillary effects are supposed to be dominant and viscous effects do not affect the liquid distribution. However, the role of liquid viscous effect cannot be neglected in our small pore networks. As a consequence, the condition of local capillary equilibrium which is one of the key assumptions in the conventional upscaling approach cannot be fulfilled. Thus, non-trivial, upgraded CMs that cannot otherwise be derived and parameterized are the target. This chapter is organized as follows: The diffusion-based CM and the CM that includes three zones defined through three individual governing equations and combined with flux boundary conditions are evaluated in Section 7.2. PNM and explanation on the method that is used to determine macroscopic parameters needed in the CM and associated results for small networks are presented in Section 7.3. In Section 7.4, the sensitivity of the macroscopic parameters on time and space is discussed. To tackle the issue of D(S) being highly associated with the process and the porous medium, we introduce a hybrid method in Section 7.5. Finally, a summary and conclusions are presented in Section 7.6.

7.2 TWO VERSIONS OF THE ONE-EQUATION CONTINUUM MODEL

In this section, we first derive the one-equation CM for drying from first principles. Then, we present this CM in a diffusion-based form (Section 7.2.1) and in a form that is called three-transport-zone CM (Section 7.2.2). Both versions of the CM can be used to describe a drying process in which the underlying transport phenomena are driven by gradients in spatially averaged quantities and controlled by non-linear parameters. Such a description is based on the hypothesis of local equilibrium within a small volume element of the drying medium, although the entire medium has not reached the equilibrium state. With air invasion, the liquid phase is divided into several clusters (dispersed phase), while the water vapor remains continuous (see Figure 7.1).

7.2.1 THE GOVERNING EQUATIONS OF DIFFUSION-BASED CONTINUUM

7.2.1.1 Model Component Mass Balance for Liquid Water

Since the liquid phase is dispersed, governing equations may not be suitable for any control volume at any time. Therefore, an essential assumption in the CM is that the liquid water remains continuous during the drying process. On this basis, the general mass balance for the liquid phase in local formulation reads

$$\frac{\partial \rho_{w,CV}}{\partial t} = -\frac{\partial}{\partial z}(\rho_{w,CV} v + j_{w,CV}) - \chi \qquad (7.1)$$

where the accumulation of liquid water in a control volume with time $\left(\dfrac{\partial \rho_{w,CV}}{\partial t}\right)$ is contributed by the convective flow $(\rho_{w,CV} v)$ and diffusion $(j_{w,CV})$ as well as the sink

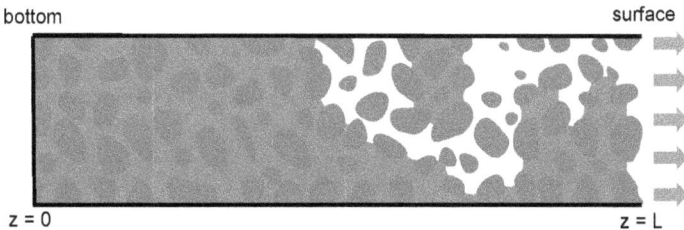

bottom   surface

z = 0   z = L

FIGURE 7.1 Schematic of the liquid water and water vapor distribution in a capillary porous medium which is open to evaporation from one side ($z=L$).

term (χ). The diffusion term can be neglected because the liquid phase contains only free water. The sink term χ is caused by evaporation during the drying process. $\rho_{w,CV}$ is the water mass density in the control volume, which can be calculated by

$$\rho_{w,CV} = \frac{m_{w,CV}}{V} \qquad (7.2)$$

Herein, $m_{w,CV}$ is the mass of liquid water in the control volume and V is the total volume. The total volume can be expressed by

$$V = \frac{V_{void}}{\varepsilon} \qquad (7.3)$$

where V_{void} is the void part of the control volume and ε is the porosity of the porous medium. The mass of liquid water in the control volume can be expressed by

$$m_{w,CV} = \rho_w V_{w,CV} \qquad (7.4)$$

where ρ_w is the mass density of liquid water and $V_{w,CV}$ is the volume of liquid water in the control volume, which can be calculated by

$$V_{w,CV} = S V_{void} \qquad (7.5)$$

Here, S is the saturation ($0 < S \leq 1$). By combining Eqs. (7.2)–(7.5), $\rho_{w,CV}$ can be determined by

$$\rho_{w,CV} = \varepsilon \rho_w S \qquad (7.6)$$

The velocity of liquid (v) can be calculated by Darcy's law (neglecting gravity):

$$v = \frac{-\dfrac{k}{\mu}\dfrac{\partial P_w}{\partial z}}{\varepsilon S} \qquad (7.7)$$

By introducing Eqs. (7.6) and (7.7) into Eq. (7.1), the mass balance for liquid water can be rearranged into

$$\frac{\partial(\varepsilon \rho_w S)}{\partial t} = -\frac{\partial}{\partial z}\left(-\rho_w \frac{k \partial P_w}{\mu \partial z}\right) - \chi \qquad (7.8)$$

where k is the effective permeability and μ is the liquid viscosity.

7.2.1.2   Component Mass Balance for Water Vapor

The local mass balance for vapor can be formulated in general as

$$\frac{\partial \rho_{v,CV}}{\partial t} = -\frac{\partial}{\partial z}(\rho_{v,CV} v + j_{v,CV}) + \chi \tag{7.9}$$

The accumulation of the vapor in the control volume ($\frac{\partial \rho_{v,\,CV}}{\partial t}$) can be determined by the convection term ($\rho_{v,\,CV} v$) and by the diffusion term of vapor in air ($j_{v,\,CV}$). χ is the source term which is equal to the sink term of liquid water at slow isothermal drying conditions. $\rho_{v,\,CV}$ is the density of water vapor in the control volume. This work assumes that drying takes place at room temperature and that total pressure within the drying material remains constant and equal to the total ambient pressure. This means that there is no gas flow in the drying material and the convection term can be neglected in Eq. (7.9).

Stefan's law is used to express the one-sided diffusion of vapor in the gas phase. Thus, the vapor flux can be calculated as

$$j_v = \frac{\tilde{M}_v \dot{N}_v}{A} = -\frac{D_{v,0}\tilde{M}_v}{\tilde{R}T}(\frac{P}{P - P_v})\frac{\partial P_v}{\partial z} \tag{7.10}$$

where $\tilde{M}_v$ is the molar mass of vapor, $D_{v,0}$ the vapor diffusion coefficient, $\tilde{R}$ the universal gas constant, P the atmospheric pressure, and P_v the vapor pressure.

Using the total cross-sectional area in place of A in Eq. (7.10) yields

$$j_{v,CV} = j_v \tag{7.11}$$

The density of vapor in the control volume ($\rho_{v,\,CV}$) can be determined by

$$\rho_{v,CV} = \frac{m_{v,CV}}{V} = \frac{\rho_v V_{v,CV}}{V} \tag{7.12}$$

with

$$V_{v,CV} = V_{void} - V_{w,CV} \tag{7.13}$$

Herein, ρ_v is the vapor density.

The density of the vapor in the control volume results in

$$\rho_{v,CV} = \varepsilon \rho_v (1 - S) \tag{7.14}$$

Finally, the mass balance for the vapor in local formulation becomes

$$\frac{\partial(\varepsilon \rho_v (1 - S))}{\partial t} = -\frac{\partial}{\partial z}\left(-\frac{D_{v,0}\tilde{M}_v}{\tilde{R}T}\left(\frac{P}{P - P_v}\right)\frac{\partial P_v}{\partial z}\right) + \chi \tag{7.15}$$

7.2.1.3 Mass Balance for Total Moisture

Assuming that the drying process is quasi-steady with regard to the vapor transfer in the gas phase, the mass balance for the total moisture (i.e. liquid and vapor) is expressed as summation of the mass balances for liquid water (Eq. 7.8) and for water vapor (Eq. 7.15):

$$\varepsilon \rho_w \frac{\partial S}{\partial t} = -\frac{\partial}{\partial z}\left(-\rho_w \frac{k \partial P_w}{\mu \partial z} - \frac{D_{v,0} \tilde{M}_v}{\tilde{R}T}\left(\frac{P}{P-P_v}\right)\frac{\partial P_v}{\partial z}\right) \qquad (7.16)$$

Equation (7.16) is the one-equation CM, which can be written in the form of a diffusion model by introducing relative humidity or the vapor pressure–saturation pressure ratio as

$$\varphi = \frac{P_v}{P_v^*} \qquad (7.17)$$

where P_v^* is the saturated vapor pressure.

Considering Eq. (7.17) and $\dfrac{\partial P_w}{\partial S}\dfrac{\partial S}{\partial z}$, Eq. (7.16) is finally simplified to

$$\varepsilon \frac{\partial S}{\partial t} = \frac{\partial}{\partial z}(D(S)\frac{\partial S}{\partial z}) \qquad (7.18)$$

where $D(S)$ (m²/s) is represented by

$$D(S) = \frac{k}{\mu}\frac{\partial P_w}{\partial S} + \frac{P_v^* D_{v,0} \tilde{M}_v}{\rho_w \tilde{R}T}(\frac{P}{P-P_v^*})\frac{\partial \varphi}{\partial S} \qquad (7.19)$$

Therefore, major parameter in the one-equation CM (Eq. 7.18) is the moisture transport coefficient $D(S)$.

7.2.1.4 Initial and Boundary Conditions

The initial condition is straightforward, namely that the network is initially filled with free water:

$$S(z = 0 \rightarrow L, t = 0) = 1 \qquad (7.20)$$

Two boundary conditions must be set: One is defined at the bottom of the porous medium (Figure 7.1) as

$$\rho_w D(S)\frac{\partial S}{\partial z}\Big|_{z=0,t} = 0 \qquad (7.21)$$

The other boundary condition is usually set at the surface of the porous medium which is connected to the free gas diffusion zone. The surface is assumed to be at the saturated vapor pressure during the first drying period or, as mentioned before, at vapor pressure which is corrected using the desorption isotherm. The vapor flow from the surface to the ambient can be considered to be one-dimensional and described by Stefan's law. Thus, the boundary condition at the surface can be defined as

$$-\rho_w D(S)\frac{\partial S}{\partial z}\bigg|_{z=L,\,t} = \frac{D_{v,\,0}\tilde{M}_v P}{\tilde{R}T\delta}\ln\left(\frac{P-P_{v\infty}}{P-\varphi_{surf}P_v^*}\right) \quad\quad (7.22)$$

where $P_{v\infty}$ denotes the vapor pressure in the bulk air and δ the thickness of the boundary layer located between the medium surface and the bulk air. $\varphi_{surf}P_v^*$ represents the vapor pressure at the surface. $\varphi_{surf}=1$ is often assumed when the surface contains liquid and φ_{surf} is determined from the desorption isotherms when saturation reduces in the porous medium along the course of drying. The resulting CM is called two-transport-zone CM.

7.2.2 THE GOVERNING EQUATIONS OF THREE-TRANSPORT-ZONE CONTINUUM MODEL

In this model, the porous medium is divided into three zones with respect to the mass transport process: The wet zone (saturated or two-phase flow zone) in the porous medium, the DZ with vapor diffusion in the porous medium (formed after the DF recedes from the surface), and the gas-side boundary layer with vapor diffusion (Figure 7.2). Thus, three governing equations are needed to describe the transport process in the respective zones. The combination of these three equations depends on the drying period considered.

7.2.2.1 The Continuum Model for the First Drying Period (Wet Surface)

Initially, mass transport in only two parts of the drying medium needs to be considered: The moisture transport in the wet zone (saturated and partially saturated zone) of the porous medium and the vapor diffusion in the gas-side boundary layer ($z=L$ to $L+\delta$) (Figure 7.2). It is assumed that the porous medium is filled with free water. Thus, the governing equation for the moisture transport in the porous medium (wet zone of the porous medium from $z=0$ to $z=DF$ in Figure 7.2) in local formulation

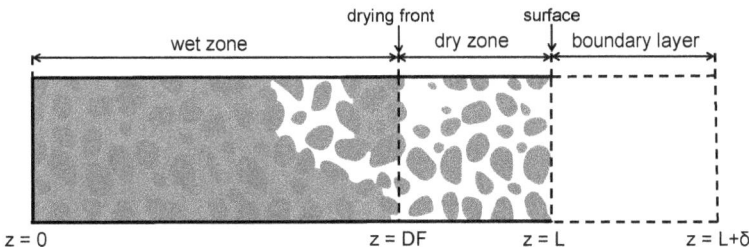

FIGURE 7.2 Three transport regions developed during the drying of a capillary porous medium.

reads as the diffusion-based equation (Eq. 7.18). Additionally, the vapor mass balance equation in the gas-side boundary layer is written as

$$\frac{\partial \varphi}{\partial t} = \frac{\partial}{\partial z}(D_{v,b}(\varphi)\frac{\partial \varphi}{\partial z}) \tag{7.23}$$

where $D_{v,b}(\varphi)$ is the vapor diffusion coefficient in the gas-side boundary layer and φ is the relative humidity defined by Eq. (7.17).

7.2.2.2   Initial and Boundary Conditions

To solve Eq. (7.23), we have to specify initial and boundary conditions. The initial condition for Eq. (7.23) should correspond to the situation when the steady state of relative humidity has been reached for a porous medium that is filled with free water. It can be calculated based on Eq. (7.23) by inserting the initial guess $(z = L \rightarrow L + \delta, \quad t = 0) = 1$, as well as the boundary conditions $(z = L,t) = 1$ to reach the steady-state distribution of the humidity, $(z = L \rightarrow L + \delta, \quad t = 0)$.

The boundary condition at the surface is

$$\rho_w D(S_{Surf})\frac{\partial S}{\partial z}\Big|_{z=L,t} = \frac{P_v^* \tilde{M}_v}{\tilde{R}T} D_{v,b}(\varphi_{surf})\frac{\partial \varphi}{\partial z}\Big|_{z=L,t} \tag{7.24}$$

The boundary conditions at the bottom of the porous medium and at the bulk of air, respectively, are

$$\rho_w D(S)\frac{\partial S}{\partial z}\Big|_{z=0,t} = 0 \tag{7.25}$$

and

$$\varphi(z = L+\delta, t) = 0 \tag{7.26}$$

7.2.2.3   The Continuum Model for the Second Drying Period (Dry Surface)

After a certain time, the liquid on the medium surface has been evaporated. Then, a DF starts receding into the porous medium. As a result, a DZ is formed between the DF and the medium surface (Figure 7.2). The interface between the foremost position of the liquid and the empty region in the porous medium is defined as the DF. The DF moves into the depth of the medium in the course of the drying process. The DF in this work is defined between two control volumes, where one is filled with saturation larger than 0.0001 and the next has saturation smaller than 0.0001. At saturation smaller than 0.0001, the control volume is assumed as empty. Thus, the position of the DF is a function of drying time, $0 < DF(t) < L$.

The governing equation reads in local formulation for vapor transport in this zone $(DF(t) \leq z < L)$:

$$\varepsilon \frac{\partial \varphi}{\partial t} = \frac{\partial}{\partial z}(D_{v,e}(\varphi)\frac{\partial \varphi}{\partial z}) \tag{7.27}$$

where $D_{v,e}(\varphi)$ (m²/s) is the vapor diffusion coefficient in this zone. Moisture transport in the wet zone $(0 \leq z < DF(t))$ can be described by Eq. (7.18). And Eq. (7.23) applies to the vapor transport in the gas-side boundary layer $(L \leq z \leq L+\delta)$.

7.2.2.4 Boundary Conditions

To couple the DZ with the wet zone in the porous medium and the gas-side boundary layer, two boundary conditions need to be specified. One boundary condition is at the DF:

$$\rho_w D(S)\frac{\partial S}{\partial z}\Big|_{z=DF,t} = \frac{P_v^* \tilde{M}_v}{\tilde{R}T} D_{v,e}(\varphi)\frac{\partial \varphi}{\partial z}\Big|_{z=DF,t} \tag{7.28}$$

and another is set at the surface of the porous medium:

$$\frac{P_v^* \tilde{M}_v}{\tilde{R}T} D_{v,e}(\varphi)\frac{\partial \varphi}{\partial z}\Big|_{z=L,t} = \frac{P_v^* \tilde{M}_v}{\tilde{R}T} D_{v,b}(\varphi)\frac{\partial \varphi}{\partial z}\Big|_{z=L,t} \tag{7.29}$$

In summary of this section, two versions of the CM have been introduced, and the respective initial and boundary conditions have been defined. To solve Eqs. (7.18), (7.23), and (7.27), three parameters are required in the form of profiles with the state variables that express $D(S)$, $D_{v,e}(\varphi)$, and $D_{v,b}(\varphi)$. Additionally, relationships between saturation (S) and relative humidity (φ) at the DF and at the surface are necessary to solve the boundary conditions for the CM. However, in the two-transport-zone CM, it only needs the state variable of $D(S)$ as well as the relationships between saturation (S) and relative humidity (φ) at the DF. Thus, in the following section, we will introduce the method used to extract the parameter profiles from simulations conducted over the limited number of realizations of PNM.

7.3 IDENTIFICATION OF THE CONTINUUM MODEL PARAMETERS FROM PNM SIMULATIONS

The algorithm for determining parameters from the PNM datasets operates in three steps. First, we virtually cut slices from the PNM. One slice comprises all throats and pores in the respective horizontal layer plus half the up and down vertical throats (Figure 7.3). Then, we calculate the mass flow rate between those slices. Every two slices are considered as the averaged volume of parameters. Finally, parameters are determined from Fick's law based on flux between slices. Moisture transport coefficients are extracted from both the TNM and the TPM.

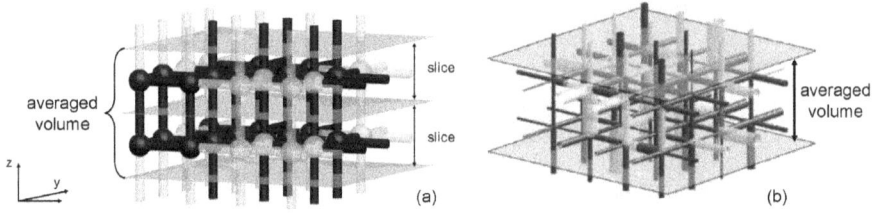

FIGURE 7.3 Schematic of two slices that are virtually cut from a throat–pore network (a) and from a throat–node network (b). Full throats/pores and empty throats/pores are shown in dark blue and light blue, respectively.

7.3.1 Moisture Transport Coefficient

First, the liquid flow rate and the vapor flow rate are computed individually by assorting, respectively, the vertically connected pores between every two slices. If both pores are filled with liquid, they are attributed to the liquid flow rate. When one of the two pores or both pores are filled with vapor, the vapor flow rate is calculated. In this way, individual transport coefficients for the liquid and for the vapor are obtained. Summing those up, the overall moisture transport coefficient and its profile upon saturation are computed.

The liquid and vapor flow rates between slices in the z-direction can numerically be calculated by summing the liquid fluxes from liquid pores (or nodes) in the nth slice to the $(n+1)^{th}$ slice:

$$j_{w,n \to n+1} = \frac{1}{A} \sum_{i,j} g_{w,ij} (P_{w,n}^i - P_{w,n+1}^j) \qquad (7.30)$$

and the vapor fluxes between the vapor pores (or nodes) in the nth slice and the $(n+1)^{th}$ slice:

$$j_{v,n \to n+1} = \frac{1}{A} \sum_{i,j} g_{v,ij} P \ln\left(\frac{P - P_{v,n+1}^j}{P - P_{v,n}^i}\right) \qquad (7.31)$$

Moisture transport coefficients for the liquid phase $(D_w(S))$ and for the vapor phase $(D_v(S))$ are calculated as

$$D_w(S) = -\frac{j_{w,n \to n+1}}{\rho_w \dfrac{S_{n+1} - S_n}{h}}, \quad D_v(S) = -\frac{j_{v,n \to n+1}}{\rho_w \dfrac{S_{n+1} - S_n}{h}} \qquad (7.32)$$

where h is the center distance between two pores. The overall moisture transport coefficient is thus obtained:

$$D(S) = D_w(S) + D_v(S) \qquad (7.33)$$

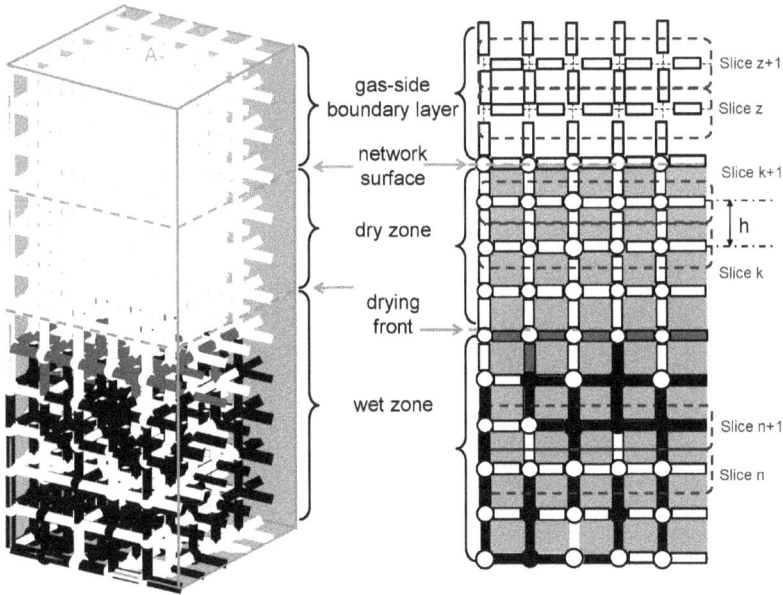

FIGURE 7.4 (left) A small throat–pore network (5×5×10) opens to evaporation from the top and (right) its front projection when $S_{net}=0.8$. A slice contains all throats and pores of one horizontal layer and half of vertical throats. The liquid and vapor throats/pores in the network are shown in dark and white, respectively. The blue throats/pores are partially filled with liquid. (Reprinted from Lu et al. (2021b).)

which is a function of local saturation (Figure 7.5):

$$S = \frac{S_{n+1} + S_n}{2} \tag{7.34}$$

Figure 7.5 shows the moisture transport coefficient for the gas phase, D_v, for the liquid phase D_w, as well as for both phases, D, obtained from the PNM drying simulations with the configuration as TNM. The data profiles are treated by averaging with the interval of network saturation, $\Delta S_{net}=0.3, 0.3, 0.4$, into three curves. The results show that these three coefficients depend on the network saturation for local saturation $S<0.68$. However, the dependence of D_w and D on network saturation is not clear for $S>0.68$. This might be a reason that the mass transport in this range is through the liquid phase only. Figure 7.6 shows the ratio of liquid flow rate (J_w) to total moisture flow rate (J), $\zeta = J_w / J$. For $S>0.68$, $J_w / J=1$, therefore the vapor flux is zero.

Figure 7.5c shows a minimum value of total moisture transport coefficient D. This minimum occurs at the saturation when Jw/J = 0.5. Thus, in each D–S curve, three critical saturation points can be observed: The limit $S \approx 1$ at which moisture transport is dominated by the liquid phase; the limit ($S \approx 0$) at which moisture transport is governed by the vapor phase; and the turning point (S_T) at which the moisture transport reaches a minimal value.

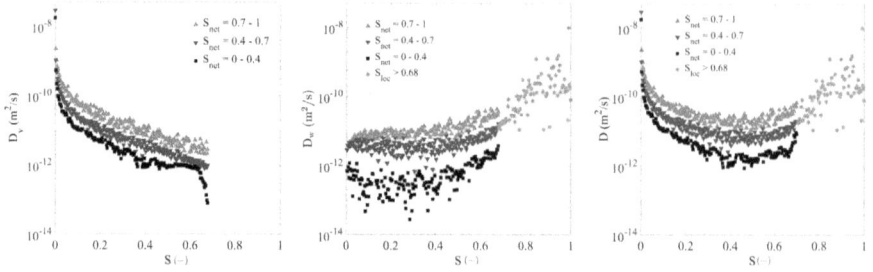

FIGURE 7.5 Moisture transport coefficients for the gas (a) and liquid (b) phases as well as the total moisture transport coefficient (c) obtained from the TNM drying simulations. The values have been averaged over local saturation intervals. (Reprinted from Attari Moghaddam et al. (2017).)

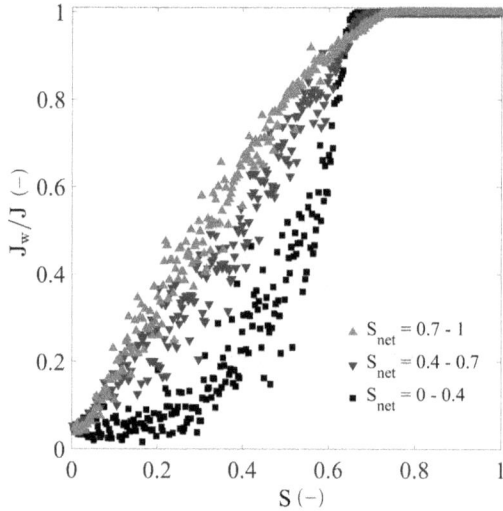

FIGURE 7.6 Ratio of the liquid flux over total flux during drying obtained from the TNM simulations at different network saturation intervals. (Reprinted from Attari Moghaddam et al. (2017).)

To reveal the transport properties in the high local saturation range, $S > 0.68$, we treated the dataset with finer intervals. The moisture transport coefficient in Figure 7.7 is obtained from the TPM, where both cylindrical throats and spherical pores contain liquid and contribute to moisture transport.

Figure 7.7 shows an additional critical saturation point that a relatively large value of D appears in the high local saturation range and it appears when slices contain the maximum of interfacial area ($S_{a_s, max}$). The interfacial area of the liquid phase (a_s) is defined as the sum of the cross-section areas of all menisci in each slice (Figure 7.8). Based on these four critical saturation points, we can subdivide the profiles of moisture transport coefficients (D–S profiles) into three regions: Liquid transport

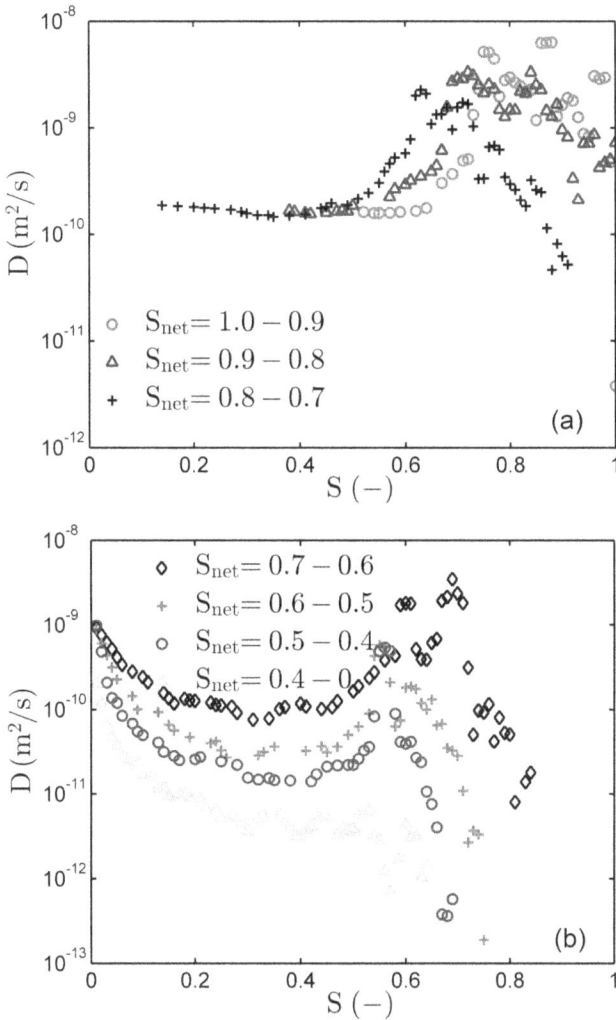

FIGURE 7.7 The moisture transport coefficient for the TPM network geometry obtained from the pore network simulations by using smaller time intervals. The values have been averaged over ten realizations for intervals of local saturation S. (Adapted from Lu et al. (2020).)

dominated region (LTDR), two-phase transport region (TPTR), and vapor transport dominated region (VTDR).

The LTDR is the range from nearly full saturation ($S \approx 1$) to $S_{as,\,max}$ in which moisture transport is fully in the liquid phase. This region can clearly be observed in Figure 7.6 where the ratio of liquid flow rate over total flow rate J_w/J is unity. At this point, the whole moisture transport occurs in the liquid phase. However, $S_{as,\,max}$ cannot clearly be seen in Figure 7.5 due to the coarse time intervals.

The dependence of the moisture transport coefficient in the LTDR on the network saturation can be seen as a consequence of the small network used here. Due to this

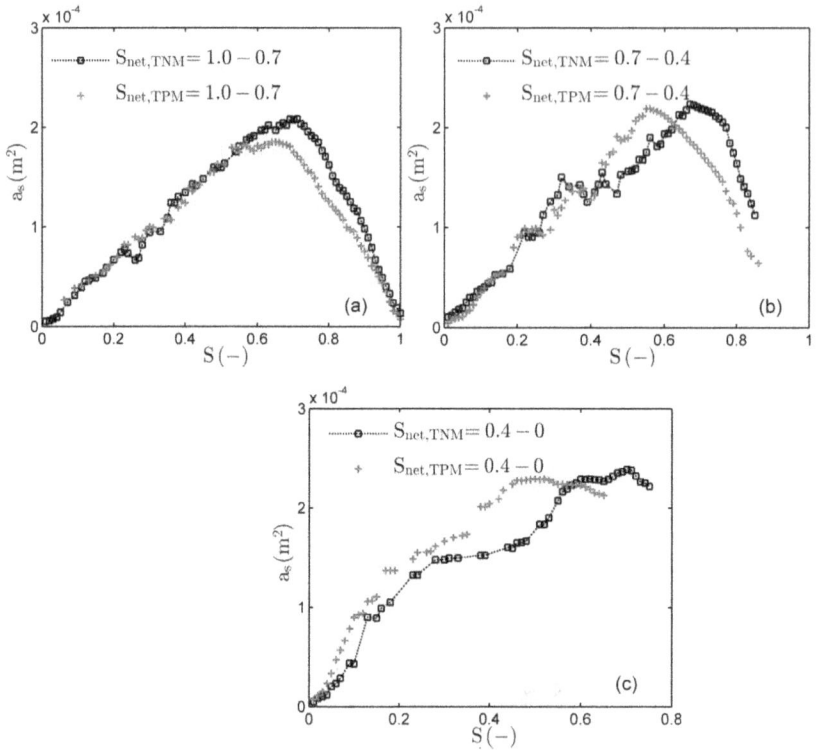

FIGURE 7.8 Interfacial area as a function of local saturation obtained from the drying simulations of TNM and TPM network geometries: (a) $S_{net}=1.0-0.7$, (b) $S_{net}=0.7-0.4$, and (c) $S_{net}=0.4-0$. The values have been averaged over ten realizations for small saturation S intervals. (Adapted from Lu et al. (2020).)

size limitation, the condition of local capillary equilibrium, which is one of the key assumptions in the conventional upscaling approach (Whitaker, 1977), cannot be fulfilled. Under this condition, capillary effects are supposed to be dominant and viscous effects do not affect the liquid distribution at the REV scale.

Slices with the maximum interfacial area ($S_{a_s, max}$) are near to the DF. The pressure drop caused by liquid viscosity can be calculated as

$$\Delta P_f = \frac{8\mu \dot{M}_w}{\pi \rho_w} \left(\frac{R_1}{R_1^4} + \frac{L_t}{R_t^4} + \frac{R_2}{R_2^4} \right) \propto \frac{H}{R^4} \tag{7.35}$$

so that it is related to the distance H between the two menisci. Long distance between two active menisci produces a high value of pressure drop. Therefore, the moisture transport coefficient increases from high saturation slices (in bottom layers with long distance for moisture transport) to slices with $S_{a_s, max}$. This could be interpreted as a consequence of the lack of length scale separation in the PNM simulations.

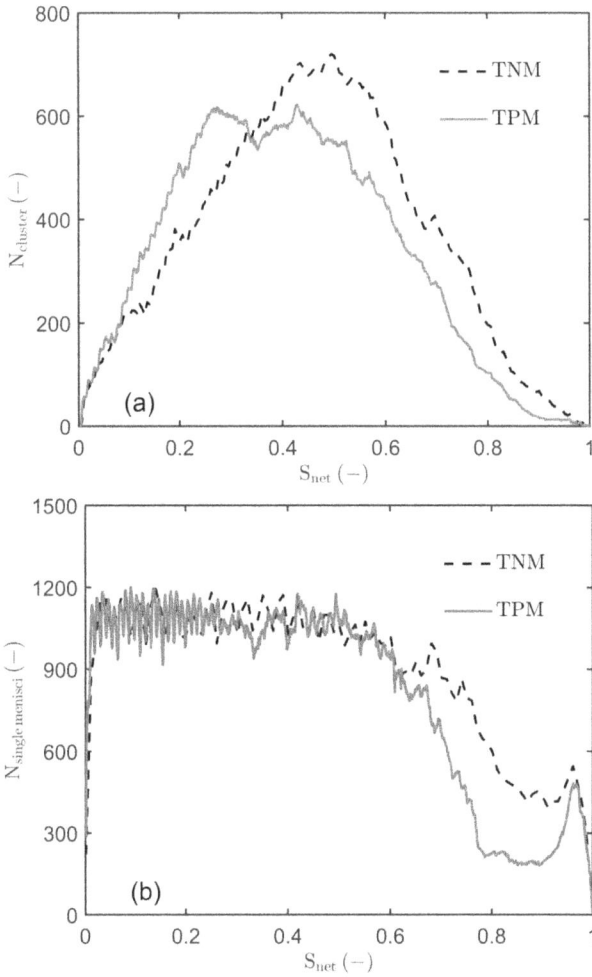

FIGURE 7.9 Numbers of (a) liquid clusters and (b) single menisci obtained from the TNM and TPM drying simulations. The values have been averaged over ten realizations for small saturation intervals. (Adapted from Lu et al. (2020).)

The TPTR is a region in which both the liquid and vapor phases contribute to moisture transport. In this region from $S_{ds,\,max}$ to S_T, the moisture transport coefficient (D) decreases by increasing internal resistance; below S_T, the moisture transport is enhanced again, this is why S_T is called the turning point. S_T can be determined according to the ratio of liquid flow rate to total moisture flow rate (Figure 7.6). When $J_w/J = 0.5$ (i.e. same contribution of liquid and vapor phases to the moisture flow), we have observed the minimum value of moisture transport coefficient (Figure 7.7). In this region, the moisture transport in the vapor phase starts to play a role. The liquid phase is broken into liquid clusters and single menisci (Figure 7.9). As a result, the moisture transport coefficient decreases from $S_{ds,\,max}$ to S_T.

The VTDR is observed for local saturation smaller than S_T where moisture transport occurs dominantly in the vapor phase. With air invasion, the liquid turns into a dispersed phase, forming many single menisci and liquid clusters (Figure 7.9). Whereas liquid viscosity controls the liquid transport mechanism, the moisture transport in the vapor phase is driven by differences in vapor pressure. The vapor pressure in the two-phase zone is related to the active interfacial areas a_s. The distribution of the interfacial area in the network leads to the uneven distribution of vapor pressure. The vapor pressure at the surface and two-phase region is expressed as the relative humidity profile.

7.3.2 RELATIVE HUMIDITY

From PNM perspective, different vapor pressures exist at the exits of network surface throats (Figure 7.10). The air is saturated with vapor (dark blue) above the partially filled and completely filled surface throats. In contrast, dry surface throats lead to unknown surface vapor pressure (light blue in Figure 7.10), which needs to be specified by the solution of the system of mass balances. We used the ratio of average vapor pressure at the surface ($P_{v,\,surf}$) from the PNM over the saturated vapor pressure (P_v^*) to obtain the surface relative humidity:

$$\varphi_{surf}(S_{surf}) = \frac{\langle P_v \rangle_{surf}}{P_v^*} \tag{7.36}$$

Moreover, we have calculated surface saturation S_{surf} from the PNM data. S_{surf} comprises all throats and pores shown in Figure 7.10: Vertical throats that end at the surface, their underlying layer of pores, and horizontal throats connecting the pores of this sub-surface layer. S_{surf} is the volume of liquid in all those pores and throats divided by their total volume.

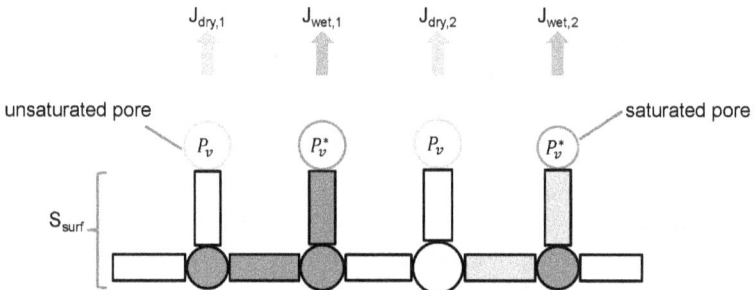

FIGURE 7.10 Schematic of local evaporation rates from the surface estimated from pore network modeling: Full throats/pores (dark blue), partially filled throats/pores (light blue), and dry throats/pores (white). Vapor pressure above wet patches (full or partially filled throats/pores) is assumed at saturation vapor pressure; vapor pressure above dry throats/pores (light blue) is unknown and needs to be determined from the local mass balances. (Adapted from Lu et al. (2021a).)

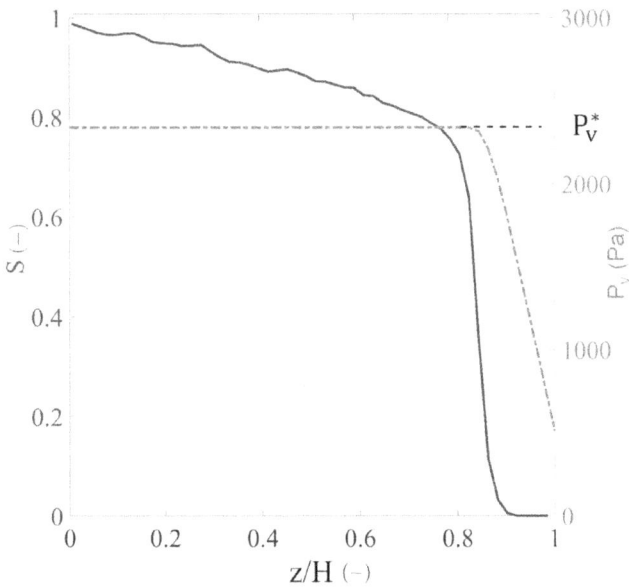

FIGURE 7.11 Profiles of the slice-averaged vapor partial pressure (P_v) and the local satura-tion (S) over the normalized network height (z/H) obtained from the TNM at network satura-tion 0.75. (Adapted from Attari Moghaddam et al. (2017).)

A similar idea is used to calculate the relative humidity profiles at the DF. In the two-phase region, the relative humidity should again be determined as a function of local saturation. This can be done by spatially averaging the partial vapor pressure of each vapor pore/node in the considered slice:

$$\varphi(S) = \frac{\langle P_v \rangle_{slice}}{P_v^*} \qquad (7.37)$$

Here, S is the average saturation in a slice containing all horizontal throats/pores of a certain layer and half of the volume of throats immediately above or below this layer (Figure 7.3).

Figure 7.11 shows the variation of the local mean vapor partial pressure and local saturation along the normalized height of the network obtained from PNM of drying when the network saturation of 0.75. The vapor pressure deviates from the saturation vapor pressure in the presence of liquid over a range of local saturation, from 0.85 to 1. In the PNM simulation, we neglected the phenomena of adsorption–desorption. Thus, the mechanism of vapor pressure in the nonzero saturation is different. From the perspective of PNM, the local thermodynamic equilibrium is satisfied at the menisci surface (where the vapor pressure above the menisci is the saturation vapor pressure) but not at the scale of the averaging volume, which is divergent from the standard CM assumptions, i.e. Whitaker, 1977. This phenomenon of vapor pres-sure with saturation in Figure 7.11 can be interpreted as the signature of a non-local equilibrium (NLE) effect.

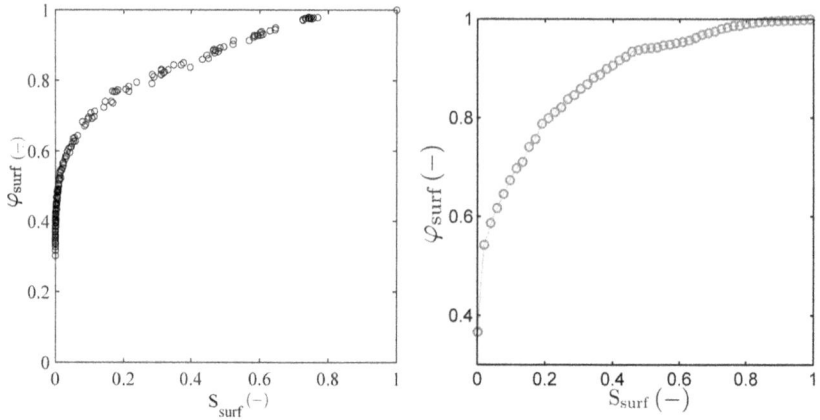

FIGURE 7.12 The surface vapor pressure–saturation relationship obtained from the TNM and TPM pore network simulations. (Reproduced from Attari Moghaddam et al. (2017) and Lu et al. (2020).)

Figure 7.12 shows the value of NLE function at the surface slice (φ_{surf}) averaged over intervals of local saturation ($\Delta S = 0.01$), namely the surface relative humidity with surface saturation. When Figure 7.12 is compared with the profile of interfacial area in Figure 7.8, we can see that the surface vapor pressure deviates from the saturation vapor pressure after $S_{as,\,max}$. The more the interfacial area formed in a slice, the more easier it is to maintain the saturated vapor pressure. Within the range of LTDR, with the air invasion, the interfacial area increases linearly to a maximum and the averaged vapor pressure in the slice is the saturation vapor pressure. Thus, $S_{as,\,max}$ is a saturation that corresponds to the relatively high moisture transport coefficient, marks the end of the LTDR, and denotes the starting point of the NLE effect.

Figure 7.13 shows the values of internal NLE function averaged over intervals of small local saturation for different values of the network saturation. It can be seen that the dependence of relative humidity on the local saturation varies during drying. Using one fitting function to express the correlation of relative humidity in the two-phase region is not enough. One can be noted that the relative humidity close to unity as the network saturation reaches zero. The reason is because the vapor diffusion path is much greater because it combines the diffusion path in the network with the diffusion path in the boundary layer, whereas at the beginning of the drying process there is only the boundary layer diffusion path.

7.3.3 Vapor Transport Coefficient

To describe vapor transport in the gas-side boundary layer and in the DZ, we need the values of vapor transport coefficient in the boundary layer, $D_{v\,b}$, and in the DZ, $D_{v,e}$.

Since we record the water and the vapor pressure distribution in the network for many intervals of network saturation (each time when network saturation has changed by $\Delta S_{net} = 0.01$), the internal moisture flow rate can be specified during

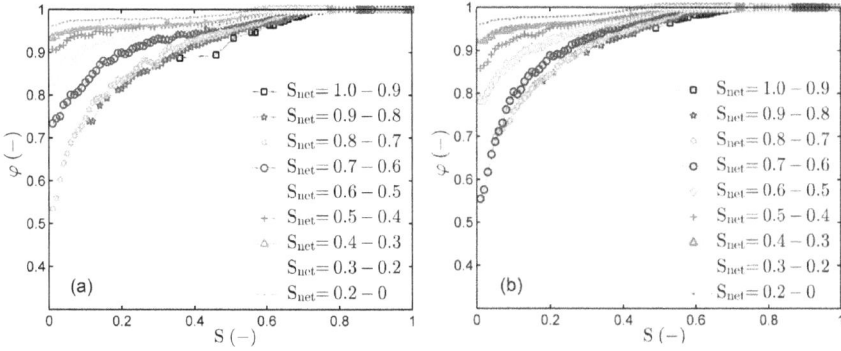

FIGURE 7.13 Mean values of internal NLE function (φ) obtained from the TNM and TPM drying simulations as a function of local saturation (S) for different network saturation (S_{net}) intervals. (Adapted from Lu et al. (2020).)

the drying process. The correlations of diffusivity with relative humidity in the DZ $D_{v,e}(\varphi)$ and in the gas-side diffusion layer $D_{v,b}(\varphi)$ can be identified based on the previously proposed method in Section 7.3.1. The basic idea is that of extracting the macroscopic parameters by averaging properties over a particular volume of the pore network. There, the averaging volume is a slice containing all horizontal throats/pores of a certain layer and half the volume of up and down throats (see Figure 7.4).

The vapor flow rate (in the z-direction) between slices can numerically be computed by summing the vapor fluxes between vapor pores in the kth slice and the $(k+1)^{th}$ slice in the DZ of the porous medium:

$$j_{v,k \to k+1,DZ} = \frac{1}{A}\sum_{i,j} g_{v,ij,DZ} P \ln\left(\frac{P - P_{v,k+1}^j}{P - P_{v,k}^i}\right) \tag{7.38}$$

where i is the vapor pore in slice k and j represents the vapor pore in slice $k+1$, $A(m^2)$ is the cross-section area of the network, and $g_{v,ij,DZ}$ is the conductance, defined as

$$g_{v,ij,DZ} = \frac{D_{va}\tilde{M}_v}{\tilde{R}T(\dfrac{R_{pi}}{A_{pi}} + \dfrac{L_t}{A_t} + \dfrac{R_{pj}}{A_{pj}})} \tag{7.39}$$

Here, A_{pi}, A_{pj}, and A_t denote the cross-sectional areas of pore i (πR_{pi}^2), A_{pi} pore j (πR_{pj}^2), and the connecting throat (πR_t^2), respectively. L_t is the throat length, which depends on the radii of the two pores (i.e. R_{pi} and R_{pj}) located at the ends of the throat with radius R_t. Similarity, the equation that holds in the gas-side boundary layer (GB) is

$$j_{v,z \to z+1,GB} = \frac{1}{A}\sum_{i,j} g_{v,ij,GB} P \ln\left(\frac{P - P_{v,n+1}^j}{P - P_{v,n}^i}\right) \tag{7.40}$$

Here, $g_{v,ij,GB}$ is considered as

$$g_{v,ij,GB} = \frac{D_{va} \tilde{M}_v A_{t,GB}}{\tilde{R}Th} \tag{7.41}$$

Since representation by pores and throats is formally continued, h(m) is the center distance between two neighbor pores in the gas-side boundary layer and $A_{t,GB}$ is the throat cross-sectional area:

$$A_{t,GB} = h^2 \tag{7.42}$$

The vapor diffusion coefficients in the gas-side boundary layer ($D_{v,\,b}\,(\varphi)$) and the DZ ($D_{v,\,e}\,(\varphi)$) obtained from

$$D_{v,b}(\varphi) = -\frac{j_{v,z \to z+1,GB}}{\dfrac{\tilde{M}_v P_v^*}{\tilde{R}T} \dfrac{\varphi_{z+1} - \varphi_z}{h}} \tag{7.43a}$$

$$D_{v,e}(\varphi) = -\frac{j_{v,k \to k+1,DZ}}{\dfrac{\tilde{M}_v P_v^*}{\tilde{R}T} \dfrac{\varphi_{k+1} - \varphi_k}{h}} \tag{7.43b}$$

can represent the vapor transport in the porous medium in a simplified way according to Fick's law that is driven by difference in vapor concentration. Vapor concentration is expressed here by φ_{n+1} and φ_n, φ_{k+1} and φ_k, φ_{z+1} and φ_z, which represent the average values of relative humidity in slices n and $n+1$, k and $k+1$, z and $z+1$, respectively. The diffusion coefficients are assigned to the arithmetic average of relative humidity between the slices:

$$\varphi = \frac{\varphi_{k+1} + \varphi_k}{2} = \frac{\varphi_{z+1} + \varphi_z}{2} \tag{7.44}$$

Figure 7.14 shows the vapor diffusion coefficient in the gas-side boundary layer against relative humidity. It should be noted that the ordinate scale in Figure 7.14 is very much magnified. This means that $D_{v,b}$ is nearly constant and nearly equal to the binary diffusion coefficient $\delta_{vg} = 2.569 \times 10^{-5}$ m^2/s at 20°C, shown as dashed line in Figure 7.14). This verifies PNM computations as well as the derivation of the parameter $D_{v,b}$. There is a miniature dependence on φ, which is attributed to Stefan flow. Equation (7.43a) does not take Stefan flow into consideration so that the influence of Stefan flow must be incorporated into the evaluated $D_{v,b}$. However, Stefan flow is small for the investigated case of slow drying (20°C), so that the dependence on φ and the deviation from the binary diffusion coefficient are both very small.

The relationship of the vapor diffusion coefficient in the gas-side boundary layer with the relative humidity can also be applied at the surface of the porous medium. Surface humidity can be determined from the surface NLE. The mass transport

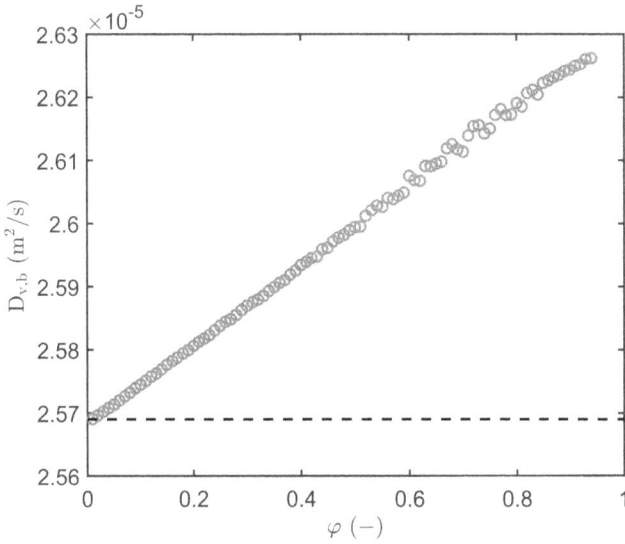

FIGURE 7.14 Vapor diffusion coefficient in the gas-side boundary layer ($D_{v,b}$) is dependent on the local relative humidity (φ). The dashed line represents the binary diffusion coefficient at 20°C. The values of $D_{v,b}$ have been averaged over ten realizations of TPM for small saturation intervals. (Adapted from Lu et al. (2021b).)

mechanism for surface humidity is the same as in the gas-side boundary layer. Thus, the correlation obtained in Figure 7.15 can be applied to the transport process in the whole region from the surface to the bulk of the surrounding gas phase.

Figure 7.15 shows the vapor transport coefficient for the DZ of the porous medium ($D_{v,e}$) over the relative humidity (φ). Identified $D_{v,e}$ is nearly constant over φ, which is again the expected behavior. It verifies PNM computations as well as the derivation of this parameter. The very weak dependence on φ may be due to Stefan flow in the throats (considered in PNM, but not in Eq. 7.43b), similar to before. With a molecular diffusion coefficient of $2.569 \times 10^{-5}\,\mathrm{m^2/s}$, a significantly lower value of $D_{v,e}$ is expected. Smaller diffusion coefficients in porous media than in the unconfined gas phase are usually attributed to obstruction (effect proportional to porosity) and tortuosity. With a total porosity of 0.35, $D_{e,v}$ by obstruction should be $8.99 \times 10^{-6}\,\mathrm{m^2/s}$, which is significantly larger than observed. This would additionally imply a tortuosity of 1.55, much larger than one would expect for a simple cubical lattice (eventually equal to 1). However, in our synthetic porous medium, with large pores at network grid points and smaller throats in between, one should rather expect the throats alone to define the obstruction effect. With 0.26 volume porosity due to the throats, $D_{v,e}$ of $6.68 \times 10^{-6}\,\mathrm{m^2/s}$ by obstruction is obtained. Considering cross-sectional porosity by the throats, which is 0.17, $D_{v,e}$ of $4.37 \times 10^{-6}\,\mathrm{m^2/s}$ is obtained. Identified $D_{v,e}$ lies with approximately $5.8 \times 10^{-6}\,\mathrm{m^2/s}$ between those values and appears very reasonable. Working with the volume porosity of throats, the remaining tortuosity effect of $6.68/5.8 = 1.15$ may well be due to the lateral (i.e. the sides) expansion of cross-section from the throats to the pores (which have by definition larger radii). In total,

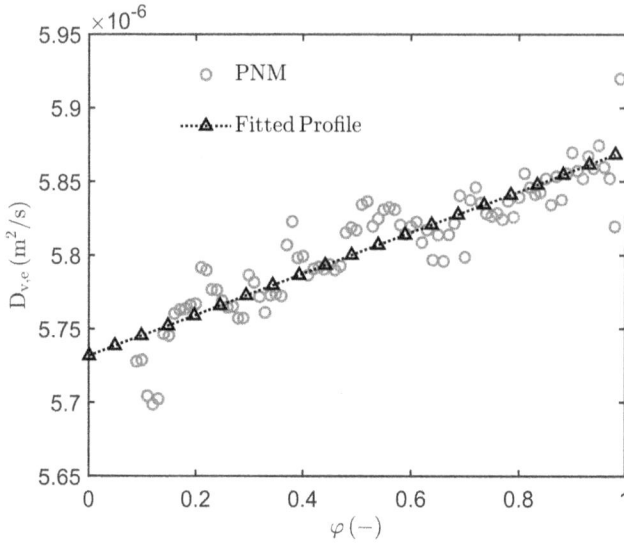

FIGURE 7.15　Vapor transport coefficient in the dry zone ($D_{v,e}$) against local relative humidity (φ). The red circles are the volume-averaged original data and the black triangles denote the fitted profile. The values of $D_{v,e}$ have been averaged over ten realizations of TPM for small saturation intervals. (Adapted from Lu et al. (2021b).)

the level of $D_{v,e}$ seems to verify our PNM computations and the conducted evaluation of this parameter. Similarly, there is a miniature dependence on φ, which is attributed to Stefan flow in Figure 7.15.

7.4　SENSITIVITY UPON THE MACROSCOPIC PARAMETERS OF CM

Moisture transport coefficients $D(S_{net}; S)$ play a prominent role among the five macroscopic parameters (Section 7.3) in the operation of the CM because of the dependence of this model parameter on the global state of the drying process (network saturation S_{net}), the position in the porous medium (local saturation S) as well as the pore structure. Figure 7.5 determined the $D(S)$ function by averaging within consecutive intervals of 0.30, 0.30, and 0.40 in network saturation from the PNM dataset. However, Figure 7.7 shows moisture transport coefficients averaged over smaller intervals of network saturation, namely $\Delta S_{net} = 0.10$. Through averaging over different intervals of network saturation, we have obtained different $D(S_{net}; S)$ profiles. Thus, the network saturation interval can affect the performance of the CM.

7.4.1　NETWORK SATURATION INTERVAL

To better understand this aspect, we performed three-transport-zone CM computations by using the macroscopic parameters as introduced in Section 7.2, but testing various $D(S_{net}; S)$ profiles obtained with different averaging intervals from the PNM dataset to predict the local saturation distribution at a high global saturation

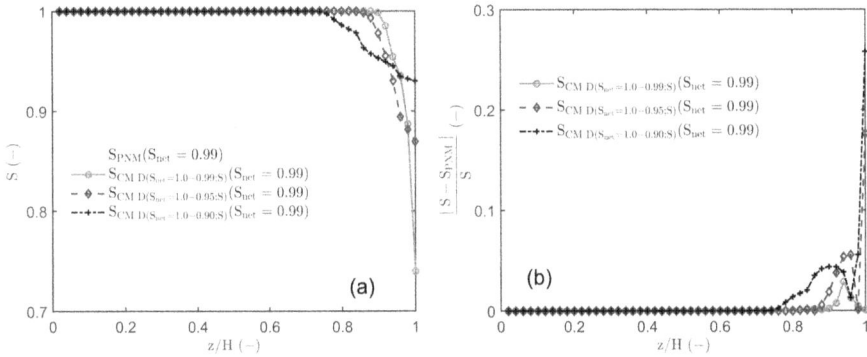

FIGURE 7.16 (a) The saturation profiles at $S_{net}=0.99$ obtained from the TPM and from the three-transport-zone CM with various $D(S_{net}; S)$ profiles, as well as (b) the normalized difference between those methods. Different averaging intervals of network saturation have been used in the determination of D, corresponding to $D(S_{net}=1.0-0.99; S)$, $D(S_{net}=1.0-0.95; S)$, and $D(S_{net}=1.0-0.90; S)$. (Adapted from Lu et al. (2021b).)

of $S_{net}=0.99$ (Figure 7.16). $D(S_{net}; S)$ has been selected for intervals of 0.01, 0.05, and 0.1 network saturation: $D(S_{net}=1.0-0.99; S)$, $D(S_{net}=1.0-0.95; S)$, and $D(S_{net}=1.0-0.90; S)$. Figure 7.16 indicates that $D(S_{net}; S)$ from the largest time interval fails to predict the saturation distribution at $S_{net}=0.99$. The closest profile to the pore network modeling results is achieved with $D(S_{net}=1.0-0.99; S)$. Figure 7.16b also shows the normalized differences of local saturations predicted by the CM from reference values of saturation (directly obtained from PNM) along with the network. Figure 7.16b demonstrates that the highest difference appears at the surface, which might affect the performance of the CM in calculating the evaporation rate. Thus, $D(S_{net}; S)$ from the smallest interval of network saturation on the PNM dataset could lead to a better prediction by means of the CM. However, it would not be practicable to provide hundreds of $D(S_{net}; S)$ profiles for the continuous modeling of drying. Therefore, the interval of network saturation used in PNM to estimate the $D(S_{net}; S)$ profiles should be at least restrained to network saturations with similar global drying characteristics.

$D(S_{net}; S)$ at high saturation is scattered both in Figures 7.5 and 7.7. Such scatter might mainly be due to the consideration of a too small network and inadequate averaging volume.

7.4.2 METHODS FOR TREATING DATASET

On one hand, the datasets obtained from the PNM simulations are discontinued points. On the other hand, the CM needs continuous profiles of parameters. One can treat the datasets by fitting into functions or can directly use the scattered datasets but interpolated linearly between each two points. Figure 7.17 shows $D(S_{net}=1.0-0.90; S)$, i.e. moisture transport coefficients gained by averaging PNM data for network saturations from 1.0 to 0.9, as well as by averaging over ten realizations. $D(S_{net}=1.0-0.90; S)$ is scattered when the local saturation is larger than the

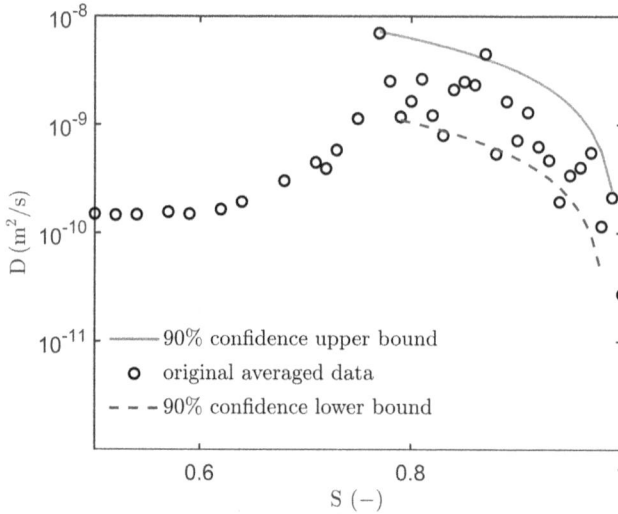

FIGURE 7.17 The moisture transport coefficient obtained from pore network TPM simulations by averaging over a small time interval of network saturation between $S_{net} = 1.0$ and $S_{net} = 0.90$ and together with 90% confidence bound. The values have been averaged over ten realizations for intervals of local saturation S. (Adapted from Lu et al. (2021b).)

irreducible saturation, $S_{irr} = 0.78$. Figure 7.17 also shows the 90% confidence bound of the moisture transport coefficient from the PNM. The 90% confidence bounds indicate that fitting of the function between these bounds would be reasonable. Thus, $S(S_{net}; z)$ profiles predicted by using different fitting functions of $D(S_{net} = 1.0 - 0.90; S)$ would be confined between extremal $S(S_{net}; z)$ functions simulated by inserting the upper and the lower bound functions into the three-transport-zone CM.

Figure 7.18 shows $S(S_{net}; z)$ profiles predicted by the three-transport-zone CM by inserting $D(S_{net} = 1.0 - 0.90; S > 0.78)$ according to either the original averaged data, or the upper bound fitting function, or the lower bound fitting function. In all cases, $D(S_{net} = 1.0 - 0.90; S < 0.78)$ is the same, and all cases are compared with the same profile from PNM results, $S_{PNM}(S_{net} = 1.0 - 0.90; z)$. Figure 7.18 indicates that the saturation distribution which is based on $D(S_{net} = 1.0 - 0.90; S > 0.78)$ from the averaged data is closer to $S_{PNM}(S_{net} = 1.0 - 0.90; z)$, compared with the $S_{CM}(S_{net} = 1.0 - 0.90; z)$ profiles from the upper and lower bound fitting functions.

Comparison of the normalized error in the prediction of local saturation in the network (Figure 7.18a) suggests that the fitting approach in the high saturation range would cause notable differences in saturation prediction, especially at the surface, where the upper bound profile evolves 100% normalized error. The poor prediction of saturation at the surface could, for example, result in the surface of the porous medium being dried earlier than expected. Thus, the relative humidity at the surface might jump out of the correlation according to surface NLE for conditions that would not correspond to PNM for the given network geometry setting. Such errors may bring troubles in boundary conditions to the CM and eventually terminate the computation so that they need to be prevented.

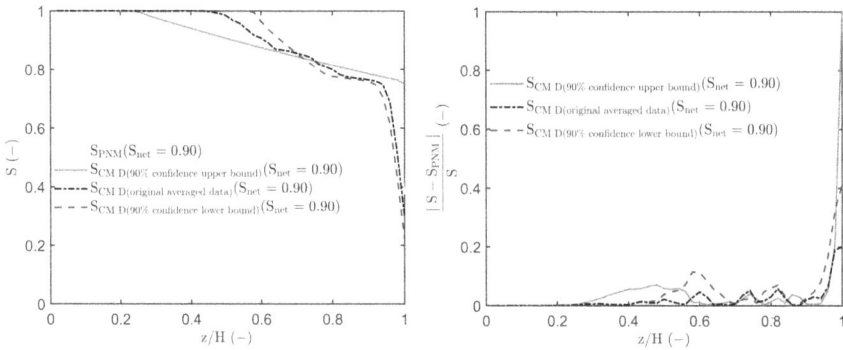

FIGURE 7.18 (a) The saturation profile at $S_{net}=0.90$ obtained from the TPM and the three-transport-zone CM with $D(S_{net}=1.0-0.90; S>0.78)$ according to the original averaged data, the upper bound fitting function, or the lower bound fitting function. Same $D(S_{net}=1.0-0.90; S<0.78)$ has been used at lower global saturation. (b) The normalized error between the PNM and CM results. (Adapted from Lu et al. (2021b).)

7.5 HYBRID METHOD TO TREAT SENSITIVITY PARAMETERS ON CM

To tackle the issue of the influence of moisture transport coefficients, which are highly associated with the network saturation, local saturation, and pore structure, on the performance of the CM, Figure 7.7 identified 7 $D(S)$ profiles, where one profile is combined for the range $S_{net}=0.4 - 0$. This has been done by defining intervals of global saturation with a width of $\Delta S_{net}=0.10$ each. Additionally, PNM data originating from ten parallel realizations have been averaged. To reduce the possibility of termination of the algorithm, the saturation distribution $S(S_{net}; z)$ obtained from the CM for each network saturation interval of 0.10 has been compared with the reference saturation. If saturation at the surface did not match, some data had to be ruled out in the high saturation region. The principle was that once the surface saturation is lower than expected, which means the liquid in the bottom is hard to transport to the surface, the value of D has to be increased in the high saturation range, otherwise the value of D is reduced. First, the total range of local saturation was divided into ten equal intervals, i.e. $S=0.90$ to 1.0. In each interval, the mean value, i.e. $\bar{D}$ $(S=0.90-1.0)$, was calculated. Then, reasonable bounds to the $D(S)$ function were set in that interval, for example by deleting values of D larger than $5\bar{D}$ and smaller than $0.01 \bar{D}$. Linear interpolation between the remaining points was used to calculate $S(S_{net}; z)$. The bounds on $D(S)$ had to be adjusted based on the comparison of $S(S_{net}; z)$ from the CM and from the PNM. In contrast, in the current work, a new hybrid method is offered to address this issue.

The hybrid method is applied for, first, a very small interval of $\Delta S_{net}=0.01$ from $S_{net}=1.0$ to 0.99, and then larger intervals of $\Delta S_{net}=0.10$ in network saturation (i.e. to $S_{net}=0.90, 0.80, 0.70$, etc.). It generates CM results by importing ten $D(S)$ profiles (some of which are exemplarily shown in Figure 7.19) combined with the

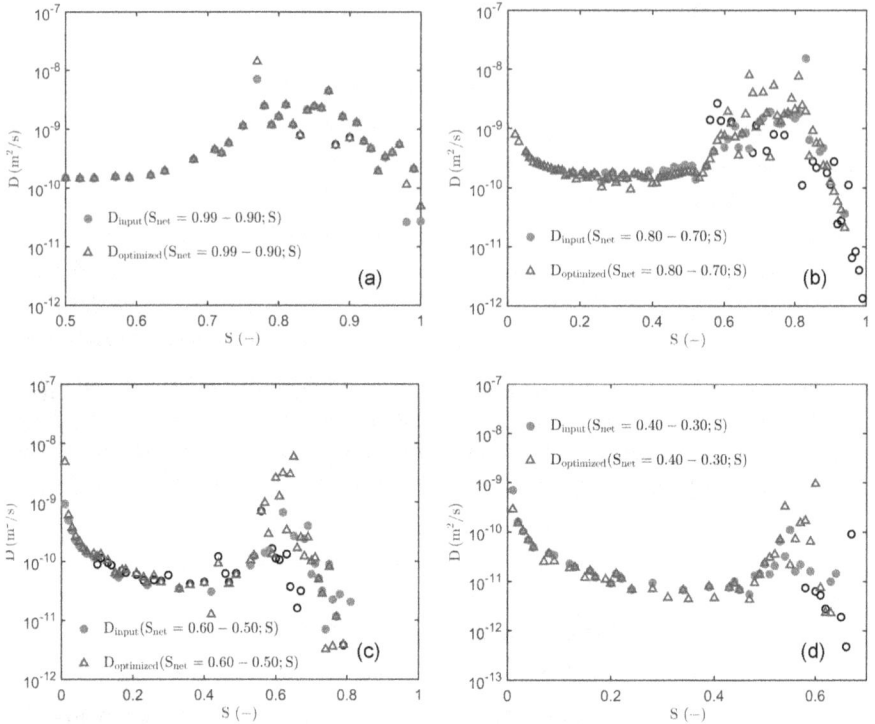

FIGURE 7.19 Moisture transport coefficients for four exemplarily shown datasets: (a) $D(S_{net}=0.99-0.9; S)$, (b) $D(S_{net}=0.8-0.7; S)$, (c) $D(S_{net}=0.6-0.5; S)$, and (d) $D(S_{net}=0.4-0.3; S)$; : data as input; Δ: optimized by the hybrid method; o: data previously deleted from the dataset. (Adapted from Lu et al. (2021b).)

other four parameters (Section 7.3) into the three-transport-zone CM. For example, to compute the network saturation distribution for $S_{net}=0.70$, the hybrid method imports the $D_{input}(S_{net}=0.80-0.70; S)$ profile shown in Figure 7.19b into the three-transport-zone CM. On this basis, the distribution of local saturation in the network, $S_{CM}(S_{net}=0.80-0.70; z_i)$, is predicted. This prediction is then compared with the local saturation distribution from the PNM, $S_{PNM}(S_{net}=0.80-0.70; z_i)$, and the comparison is quantified by the objective function (E):

$$E = \frac{1}{2}\sum_i (S_{CM}(S_{net}; z_i) - S_{PNM}(S_{net}; z_i))^2 \qquad (7.45)$$

The purpose of the hybrid method is to reduce the value of the objective function. The hybrid method examines the discrepancy of the saturation for each control volume (CV) in the CM. For example, the accumulation of moisture in the ith CV is controlled by the value of the moisture transport coefficient at input and output. $S_{CM}(S_{net}; z_i) > S_{PNM}(S_{net}; z_i)$ means that the saturation of the ith CV is overestimated by the CM. To reduce the saturation at z_i, we could increase the value of D at saturation

$(S_{CM}(z_i)+S_{CM}(z_{i+1}))/2$ (output flow) and reduce D at saturation $(S_{CM}(z_i)+S_{CM}(z_{i-1}))/2$ (input flow) by a small quantity (i.e. to $(1\pm0.01)D$). The situation of $S_{CM}(S_{net};$ $z_i)<S_{PNM}(S_{net}; z_i)$ is treated inversely, however. It should be pointed out that the first CV only considers $D(S_{net}; z_i=2)$ at the output side, whereas the CV at the DF alters $D(S_{net}; z_i=DF-1)$ at the input side. Then, the adjusted $D(S_{net}; z_i)$ is an input parameter to solve the CM again. The simulation result of, for example, $S_{CM}(S_{net}=0.80-0.70;$ $z_i)$ has to be compared with $S_{PNM}(S_{net}=0.80-0.70; z_i)$ by recalculating the objective function, Eq. (7.45). The procedure will be repeated until the value of the objective function has reached a certain tolerance, i.e. 0.05. The whole process will be repeated until the network saturation $S_{net}=0.10$. Finally, ten optimized datasets $D_{optimized}(S_{net};$ $S)$ are obtained and four of them are exemplarily shown in Figure 7.19 (denoted by the blue triangles). Datasets presented in Figure 7.19 are $D(S_{net}=0.99-0.90; S)$, $D(S_{net}=0.80-0.70; S)$, $D(S_{net}=0.60-0.50; S)$, and $D(S_{net}=0.40-0.30; S)$. It should be noted that $D(S_{net}=1.0-0.99; S)$ has been averaged with $D(S_{net}=0.99-0.90; S)$, as well as $D(S_{net}=0.40-0.30; S)$, $D(S_{net}=0.30-0.20; S)$, and $D(S_{net}=0.20-0.10; S)$ have been averaged into, respectively, one profile in Figure 7.7.

Figure 7.19 also shows the moisture transport coefficients (denoted by red cycles) which have been used as input values for the hybrid method. In the four plots, also points that have previously been deleted from the dataset appear as circles. Not shown profiles (the other six profiles) are similar. The optimized $D(S_{net}; S)$ indicate strong values of the moisture transport coefficient at specific local saturations, which correspond to slices that contain maximum interfacial area. For example, the maximum of the dataset $(S_{net}=0.80-0.70; S)$ is found at ca. $S=0.70$. This value would not be easier to recognize by averaging over S_{net}. The trend agrees with the theory of governing liquid transport in the capillary–viscous regime that has been discussed in Section 7.3.

7.6   CMS VERSUS PNMS

In this section, we feed the macroscopic parameters determined in Section 7.3 or optimized $D(S)$ in Section 7.5 into two versions of CM, i.e. the two-transport-zone parameterized CM and the new three-transport-zone CM, described in Section 7.2. In the following part, we compare the overall CM prediction from the two-transport-zone CM by using the fitting function with results using the original averaged data. Then, simulation results from the three-transport-zone CM by adopting the original averaged data are presented and compared with the simulations obtained from the two-transport-zone CM.

7.6.1   Simulation Results Obtained from the Two-Transport-Zone CM

To solve the two-transport-zone CM (Eq. 7.18), we need the moisture transport coefficient D and the NLE functions, φ. Figure 7.5 shows the three curves of $D(S)$ and Figure 7.12a is the surface NLE obtained from TNM simulations. When using these two parameters to solve Eq. (7.18), one may face the issue that the local saturation reaches very small values for which no values of D and φ_{surf} are available. This is due to the fact that local saturation can be captured only until a certain positive value; therefore, the values of these two parameters are unknown for very small local saturation. A solution may be to increase the network size, so that the aforementioned parameters can be calculated for lower local saturation. This solution however increases the computation time and might not solve the problem completely since the local saturation in the CM could still become lower than the smallest saturation of the slice.

In order to obtain values of D at very low local saturation, the following method based on the evaporation rate obtained by PNM drying simulations is used: After fitting a curve to the available data points of the moisture transport coefficient and NLE function at the surface φ_{surf}, the discretized one-equation continuum drying model is solved. When the surface saturation reaches very small values, for which D and φ_{surf} values are not available ($S_{surf} < 2 \times 10^{-4}$), the averaged evaporation rate from the PNM drying simulations is applied as the boundary condition. Then, the value of D for the corresponding surface saturation is obtained, provided that the mass conservation at the surface is satisfied. In this way, the moisture transport coefficient of any other node with $S_{surf} < 2 \times 10^{-4}$ can be obtained. Then, a curve is fitted through all data points of D. By applying the same procedure for the evaporation rate (using φ_{surf} to calculate the evaporation rate for $S_{surf} > 2 \times 10^{-4}$ and the averaged evaporation rate for the PNM drying simulations for $S_{surf} < 2 \times 10^{-4}$) and using the new fitted curve for D, this time the two-transport-zone CM can be solved until the network dries completely.

The normalized averaged evaporation rates computed by the two-transport-zone CM and PNM drying simulations are shown in Figure 7.20. The evaporation rate is computed with the two-transport-zone CM as long as data points of φ_{surf} are available ($S_{net} \sim 0.85$). Afterward, the evaporation rate of PN drying simulation is used as boundary condition in the CM (Eq. 7.22). As can be seen, the evaporation rate curves resemble each other very well. This of course only shows that the two-transport-zone CM model can reproduce the data and is by no means a proof that the two-transport-zone CM model can predict the evaporation rate. This is a reason that the boundary condition is fed with the simulation results from TNM drying simulations. The saturation profiles obtained from the two methods are plotted against the network depth in Figure 7.21. As can be seen, the CM is able to reproduce the saturation profiles of the TNM simulations for drying over the full saturation range. However, there is a price that has to be paid for this agreement, namely that the local relative humidity of air has to be fitted as a function of the local saturation for both the interior of the material and its surface.

Instead of using the fitting method, the alternative is to directly apply averaged datasets, for example Figure 7.7, into the two-transport-zone CM. The linear regression is used to assess values between every two data. Another issue is that the previous approach into the two-transport-zone CM is not satisfied as the drying rate is the expected outcome, not as the feeding parameter. This could be solved by considering the boundary condition (Eq. 7.22) at the development of the DF. Once the surface recedes, the relative humidity at the DF, φ_{DF}, can be determined from the internal NLE function. At the same time, the distance from the boundary condition to the bulk increases correspondingly.

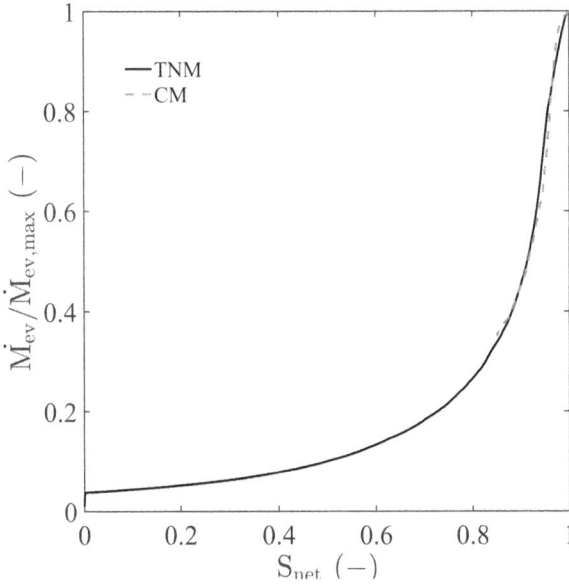

FIGURE 7.20 Normalized averaged evaporation rates obtained from the two-transport-zone CM and TNM simulations. (Adapted from Attari Moghaddam et al. (2017).)

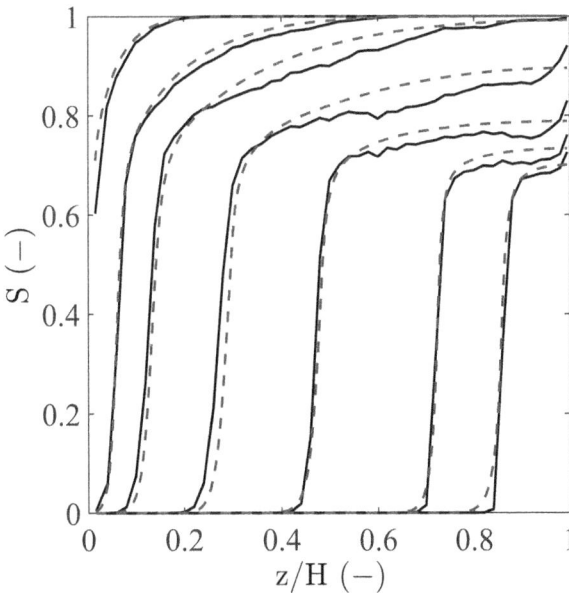

FIGURE 7.21 Saturation profiles obtained from two-transport-zone CM (blue dashed lines) and TNM (black solid lines) simulations. The saturation profiles of TNM are averaged over 15 realizations. From top, the profiles belong to network saturation of 0.98, 0.9, 0.8, 0.6, 0.4, 0.2, and 0.1, respectively. (Adapted from Attari Moghaddam et al. (2017).)

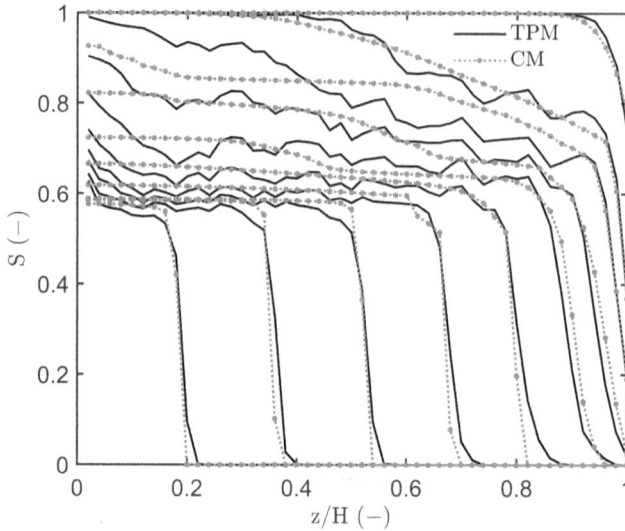

FIGURE 7.22 Saturation profiles obtained from the dynamic continuum model (red dots) and the pore network simulations for the throat–pore configuration (black lines). TPM saturation profiles are averaged over ten realizations. From top, the profiles belong to network saturations of 0.99, 0.90, 0.80, 0.70, 0.60, 0.50, 0.40, 0.30, 0.20, and 0.10, respectively. (Adapted from Lu et al. (2020).)

Figure 7.22 shows the saturation profiles obtained from the pore network simulations with the TPM configuration and the solution of the CM. The two-transport-zone CM predicts fairly the saturation profiles, except those profiles at high network saturation where the numerical values of the moisture transport coefficient are scattered (see Figure 7.7). A fairly good agreement of saturation at the DF is observed. By comparing Figure 7.22 with Figure 7.21, a better agreement is observed at the DF.

However, we observed that it is still hard to provide a good agreement for the high saturation slices especially at the network bottom (Figures 7.21 and 7.22). Another view is that this is an edge effect whose size is smaller than a REV size (Talbi and Prat, 2021). Accordingly, a standard CM is not supposed to capture this effect which is of negligible relative extension in large pore networks. Moisture transport coefficient for low local saturation slices ($S < 2 \times 10^{-4}$) is decided according to the evaporation rate from the pore network in Figure 7.21. However, Figure 7.22 considers the Stefan flow to the region from the DF to the gas bulk, while the DF traverses the network. The vapor pressure is defined according to the value of φ within and at the surface of the network. Since the empty region in the porous medium is added into the gas-side diffusion layer, it leads to the fair agreement at the high saturation because the moisture transport coefficient in the empty region is smaller than that in the free space. Thus, the one-equation CM does not meet the demand to describe the drying process.

7.6.2 SIMULATION RESULTS OBTAINED FROM THE THREE-TRANSPORT-ZONE CM

To operate the three-transport-zone CM, the values of five parameters should be known: The moisture transport coefficient $D(S)$, the vapor diffusion coefficient in the DZ $D_{v,e}(\varphi)$, the vapor diffusion coefficient in the gas-side boundary layer $D_{v,b}(\varphi)$; moreover, the NLE function at the surface $\varphi_{surf}(S_{surf})$ and at the DF $\varphi_{DF}(S_{DF})$. Here, the moisture transport coefficient $D(S)$ is selected from the optimized $D(S)$ computed by the hybrid method. The rest of parameters can be found in Section 7.3.

Figure 7.23 shows the drying kinetics and saturation profiles obtained by the three-transport-zone CM combined with $D(S_{net}; S)$ from the hybrid method, as well as compared with the results obtained from the TPM. As can be seen, the macroscopic CM is capable of reproducing the drying characteristics of capillary porous media.

The agreement between the adequately parameterized CM and the PNM is strong, especially after a DF has started receding from the surface into the medium. The saturation profiles obtained from the three-transport-zone CM demonstrate the following: The edge effect that appears as a sharp drop in saturation at positions near the evaporative surface in the initial drying period (see, e.g., Le Bray and Prat, 1999; Faure and Coussot, 2010; Or et al., 2013), fluctuations in the saturation distribution which are caused by the random pore size distribution, and the viscosity effect that holds the liquid at the bottom of the porous medium (Figure 7.23b). These effects are overlooked by the classical approach (Pel et al., 2002; Pel and Landman, 2004).

Discrepancy in the prediction of the diffusion-based CM is shown in Figure 7.22 and it is compared to the respective discrepancy of the CM according to Figure 7.23. This is performed in two ways: (a) Calculating the performance errors by summing the discrepancy for the entire drying period along the length of the network and (b) identifying the performance errors by summing the discrepancy over the entire length of the network for each of 0.1 network saturation.

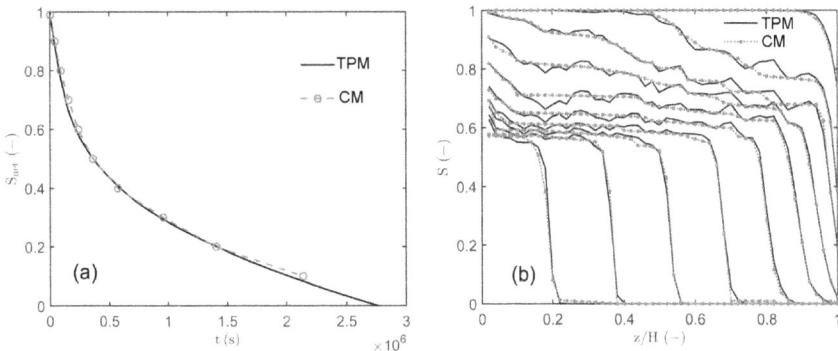

FIGURE 7.23 (a) Drying kinetics and (b) saturation profiles obtained from the three-transport-zone CM and the TPM simulations. The saturation profiles of the TPM are averaged over ten realizations. From top, the profiles belong to network saturations of 0.99, 0.90, 0.80, 0.70, 0.60, 0.50, 0.40, 0.30, 0.20, and 0.10, respectively. (Adapted from Lu et al. (2021b).)

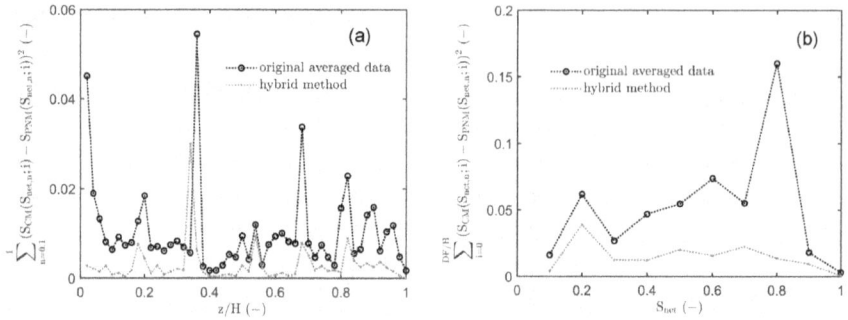

FIGURE 7.24 Deviation from PNM results of predictions of CM using the hybrid method in three-transport-zone CM vs. two-transport-zone CM with original averaged parameter identification in Figure 7.7: (a) performance errors by summing the discrepancy over the entire drying period along the length of the network and (b) performance errors by summing the discrepancy over the entire length of the network for each 0.10 of network saturation. (Adapted from Lu et al. (2021b).)

Figure 7.24a indicates that three-transport-zone CM associated with optimized $D(S_{net}; S)$ from the hybrid method provides a better prediction not only at the surface but also among the bottom slices. Additionally, the three-transport-zone CM, coupled with the hybrid method, shows a strong agreement with the TPM along the whole drying period (Figure 7.24b). The highest improvement has been seen for the network saturation profile at $S_{net} = 0.80$. However, some errors in saturation prediction using the hybrid method can still be observed in Figure 7.24. One reason may be that the amount by which the moisture transport coefficient is changed within its step is arbitrary and fixed in the hybrid method, i.e. 0.01 of the previous value of D is assumed in the current work.

Nevertheless, it can be seen that the dataset from the hybrid method preserves much of the transport information from the primary dataset (created by PNM) when used in the frame of the CM. Since the saturation profile reflects the complex interplay of capillary and viscous forces and the structural impact, it can be said that these interactions are preserved by the scattered values of moisture transport coefficient that the hybrid method provides. Contrary, such information is at least partially destroyed by the fitting approach to moisture transport coefficients, i.e. when extensively averaged and smooth functions are used to approximate this parameter.

7.7   SUMMARY AND CONCLUSIONS

In this work, we revisited the one-dimensional CM of drying by means of extensive numerical experiments performed using two specific classes of three-dimensional discrete PNMs called TNM and TPM. Datasets generated by these PNMs served as benchmark reference solutions and they were also used to compute the macroscopic parameters of the CM after postprocessing. Two situations were distinguished based on boundary conditions applied to solve the CM. When the evaporation rate at the porous medium surface is set as a boundary condition, the CM is called two-transport-zone CM. The evaporation rate here is obtained from the PNM simulations. This, however, limits the application of the CM since the evaporation rate is one expected outcome from the CM. Flux boundary conditions imposed at the DF which evolve freely during drying were another way to solve the CM. The resulting CM is called three-transport-zone CM. Three transport zones correspond to the wet and dry zones in the porous medium as well as the zone at the gas diffusion boundary layer above the porous medium. Thus, three separate governing equations are used to express the mass transport process.

The two-transport-zone CM requires the moisture transport coefficient of the porous medium as well as the relationship between saturation and relative humidity at the DF. Parameters emerging in the three-transport-zone CM are the moisture transport coefficient in the wet zone within the porous medium, the vapor transport coefficient in the DZ within the porous medium, the vapor transport coefficient in the gas-side diffusion layer, the vapor pressure–saturation relationship at the surface, as well as the vapor pressure–saturation relationship at the DF.

In this chapter, parameters are extracted from the simulations conducted with small pore networks. The algorithm in determining these parameters from PNM datasets works as follows. First, a pore network is virtually cut into slices. The slice here is considered as all throats and pores in the horizontal layer plus half the up and down vertically throats. Then, the mass flow rate between those slices is computed. Based on fluxes between slices, parameters are finally computed.

The moisture transport coefficient profiles obtained from PNM simulations are divided into three regions within the limits of full and zero saturation which are separated by two critical points. From full saturation ($S = 1$) to the saturation of maximum interfacial area ($S_{as, max}$), the LTDR is located in which moisture transport takes place by the liquid phase. From $S_{as, max}$ to S_T (saturation when $\zeta = 0.5$) is the two-phase transport region (TPTR), in which the moisture transport occurs through both the liquid and vapor phases, resisted by liquid clusters and single menisci. From S_T to the empty slice ($S = 0$) is the VTDR where the vapor phase plays a dominant role in moisture transport. Results on the relationship between the vapor partial pressure and the saturation vapor pressure within the network show a NLE effect that becomes apparent in both network geometries after $S_{as, max}$. However, the surface NLE appears at the beginning of the drying process.

The vapor transport parameters in the DZ and the gas-side boundary layer have been calculated from the PNM and shown to be nearly constant with relative humidity. The weak dependence of the vapor transport parameters on the relative humidity has been attributed to Stefan flow, which has not been taken into account in the derivation of these model parameters from PNM results. However, the influence of Stefan flow is small for the investigated case of slow drying.

After obtaining the macroscopic parameters from PNM datasets, the next step was to feed them into the CM. In this regard, one should be aware of issues resulting from the fact that the parameters are highly associated with the dynamics of the drying process and with the microstructure of the porous medium. Moreover, the averaging interval used to obtain the macroscopic parameters from the PNM plays a significant role in the performance of the CM. Therefore, averaging should be performed only within periods with similar drying behavior. The direct usage of the moisture transport coefficients extracted from the PNM shows a better performance of the CM than using the fitting approach to transfer the dataset from the PNM into the CM. In general, inappropriate values of the moisture transport coefficient may not only reduce the accuracy of CM results but even cause the termination of the CM algorithm.

To solve the issue of the CM being highly sensitive to the moisture transport coefficient, we have introduced a hybrid method to control the performance of the CM by altering locally this macroscopic parameter within a limited range. In this way, a stronger agreement between the continuum and the discrete model is observed in terms of both drying kinetics and saturation profiles. The saturation profiles predicted by the proposed CM using the macroscopic parameters from PNM indicate that the effect of the random pore size distribution, the influence of liquid viscosity, and the edge effect are properly captured. This is hardly possible to achieve by means of classical approaches to the modeling of drying processes. Effective overall vapor diffusivity has also been derived from pore network results to show that it is equal to the molecular diffusivity of water vapor in the boundary layer. Very small deviations have been observed by evaluating for equimolar diffusion pore network simulations in which the Stefan effect (one-side diffusion) has been considered. It is shown that profiles of the effective overall vapor diffusivity obtained in this way are non-linear with surface saturation but linear with relative humidity.

Next research steps shall be to use discrete pore network simulations as a guide and a template to develop and validate sophisticated CMs, such as those that account for the NLE nature of the drying process in capillary porous media.

ACKNOWLEDGMENT

This work is funded by the Deutsche Forschungsgemeinschaft (DFG, German Research Foundation) – Project-ID 422037413 – TRR 287, and the Graduate School 1554 "Micro-Macro-Interactions in Structured Media and Particulate Systems".

REFERENCES

Attari Moghaddam, A., 2017. Parameter estimation and assessment of continuum models of drying on the basis of pore network simulations. PhD thesis, Otto von Guericke University Magdeburg.

Attari Moghaddam, A., Prat, M., Tsotsas, E., Kharaghani, A., 2017. Evaporation in capillary porous media at the perfect piston-like Invasion limit: Evidence of nonlocal equilibrium effects. *Water Resources Research* 53: 10433–10449.

Bird, R.B., Stewart, W.E., Lightfoot, E.N., 2007. *Transport Phenomena*. John Wiley & Sons, Inc, New York.

Faure, P., Coussot, P., 2010. Drying of a model soil. *Physical Review E* 82: 036303.

Geoffroy, S., Prat, M., 2014. A review of drying theory and modelling approaches. In: J.M.P.Q. Delgado (ed.), *Drying and Wetting of Building Materials and Components*, 145–173. Springer International Publishing, Cham.

Gómez, I., Sala, J.M., Millán, J.A., 2007. Characterization of moisture transport properties for lightened clay brick — comparison between two manufacturers. *Journal of Building Physics* 31: 179–194.

Kharaghani, A., 2021. Drying and wetting of capillary porous materials: Insights from imaging and physics-based modeling. Habilitation thesis, Otto von Guericke University Magdeburg.

Kharaghani, A., Mahmood, H.T., Wang, Y.J., Tsotsas, E., 2021. Three-dimensional visualization and modeling of capillary liquid rings observed during drying of dense particle packings. *International Journal of Heat and Mass Transfer* 177: 121505.

Le Bray, Y., Prat, M., 1999. Three-dimensional pore network simulation of drying in capillary porous media. *International Journal of Heat and Mass Transfer* 42: 4207–4224.

Lu, X., Kharaghani, A., Tsotsas, E., 2020. Transport parameters of macroscopic continuum model determined from discrete pore network simulations of drying porous media: Throat-node vs. *throat-pore configurations. Chemical Engineering Science* 223: 115723.

Lu, X., 2021. From microscale to macroscale modeling of drying porous media. PhD thesis, Otto von Guericke University Magdeburg.

Lu, X., Kharaghani, A., Tsotsas, E., 2020. Transport parameters of macroscopic continuum model determined from discrete pore network simulations of drying porous media: Throat-node vs. *throat-pore configurations. Chemical Engineering Science* 223: 115723.

Lu, X., Tsotsas, E., Kharaghani, A., 2021a. Insights into evaporation from the surface of capillary porous media gained by discrete pore network simulations. *International Journal of Heat and Mass Transfer* 168: 120877.

Lu, X., Tsotsas, E., Kharaghani, A., 2021b. Drying of capillary porous media simulated by coupling of continuum-scale and micro-scale models. *International Journal of Multiphase Flow* 140: 103654.

Maier, L., Scherle, M., Hopp-Hirschler, M., Nieken, U., 2021. Effective transport parameters of porous media from 2D microstructure images. *International Journal of Heat and Mass Transfer* 175: 121371.

Metzger, T., Irawan, A., Tsotsas, E., 2007. Influence of pore structure on drying kinetics: A pore network study. *AIChE Journal* 53: 3029–3041.

Metzger, T., Tsotsas, E., 2008. Viscous stabilization of drying front: Three-dimensional pore network simulations. *Chemical Engineering Research & Design* 86: 739–744.

Maalal, O., Prat, M., Lasseux, D., 2021. Pore network model of drying with Kelvin effect. *Physics of Fluids* 33(2): 027103.

Or, D., Lehmann, P., Shahraeeni, E., Shokri, N., 2013. Advances in soil evaporation physics—A review. *Vadose Zone Journal* 12: 1–16.

Panda, D., Bhaskaran, S., Paliwal, S., Kharaghani, A., Tsotsas, E., Surasani, V.K., 2020. Pore-scale physics of drying porous media revealed by Lattice Boltzmann simulations. *Drying Technology* 40: 1114–1129.

Pel, L., Brocken, H., Kopinga, K., 1996. Determination of moisture diffusivity in porous media using moisture concentration profiles. *International Journal of Heat and Mass Transfer* 39: 1273–1280.

Pel, L., Ketelaarss, A.A.J., Adan, O.C.G., Van Well, A.A., 1993. Determination of moisture diffusivity in porous media using scanning neutron radiography. *International Journal of Heat and Mass Transfer* 36: 1261–1267.

Pel, L., Landman, K.A., 2004. A sharp drying front model. *Drying Technology* 22: 637–647.

Pel, L., Landman, K.A., Kaasschieter, E.F., 2002. Analytic solution for the non-linear drying problem. *International Journal of Heat and Mass Transfer* 45: 3173–3180.

Philip, J.R., Vries, D.A.D., 1957. Moisture movement in porous materials under temperature gradients. *Eos, Transactions American Geophysical Union* 38: 222–232.

Prat, M., 1993. Percolation model of drying under isothermal conditions in porous media. *International Journal of Multiphase Flow* 19: 691–704.

Prat, M., 1998. Discrete models of liquid-vapour phase change phenomena in porous media. *Revue Generale De Thermique* 37: 954–961.

Talbi, M., Prat, M., 2021. Coupling between internal and external mass transfer during stage-1 evaporation in capillary porous media: Interfacial resistance approach. *Physical Review E* 104: 055102.

Vorhauer, N., Wang, Y.J., Kharaghani, A., Tsotsas, E., Prat, M., 2015. Drying with formation of capillary rings in a model porous medium. *Transport in Porous Media* 110: 197–223.

Wang, Y., Kharaghani, A., Metzger, T., Tsotsas, E., 2012. Pore network drying model for particle aggregates: Assessment by X-ray microtomography. *Drying Technology* 30: 1800–1809.

Whitaker, S., 1977. *Simultaneous Heat, Mass, and Momentum Transfer in Porous Media: A Theory of Drying, Advances in Heat Transfer.* Elsevier. 119–203.

Wilkinson, D., Willemsen, J.F., 1983. Invasion percolation: A new form of percolation theory. *Journal of Physics A: Mathematical and General* 16: 3365.

Wu, R., Zhang, T., Ye, C., Zhao, C.Y., Tsotsas, E., Kharaghani, A., 2020. Pore network model of evaporation in porous media with continuous and discontinuous corner films. *Physical Review Fluids* 5: 014307.

Yiotis, A.G., Stubos, A.K., Boudouvis, A.G., Yortsos, Y.C., 2001. A 2-D pore-network model of the drying of single-component liquids in porous media. *Advances in Water Resources* 24: 439–460.

Yiotis, A.G., Salin, D., Yortsos, Y.C., 2015. Pore network modeling of drying processes in macroporous materials: Effects of gravity, mass boundary layer and pore microstructure. *Transport in Porous Media* 110: 175–196.

Zachariah, G.T., Panda, D., Surasani, V.K., 2019. Lattice Boltzmann simulations for invasion patterns during drying of capillary porous media. *Chemical Engineering Science* 196: 310–323.

Index

For Product Safety Concerns and Information please contact our EU
representative GPSR@taylorandfrancis.com
Taylor & Francis Verlag GmbH, Kaufingerstraße 24, 80331 München, Germany